Worzel
4
the
World

FO

QUITE actual

Very

LOVE of

Worzel Wooface

Catherine Pickles

Hubble&Hattie

The Hubble & Hattie imprint was launched in 2009 and is named in memory of two very special Westies owned by Veloce's proprietors.

Since the first book, many more have been added to the list, all with the same underlying objective: to be of real benefit to the species they cover, at the same time promoting compassion, understanding and respect between all animals (including human ones!)

Hubble & Hattie is the home of a range of books that cover all-things animal, produced to the same high quality of content and presentation as our motoring books, and offering the same great value for money.

More great Hubble & Hattie books!

Special thanks for use of their images to Kerry Jordan from Whippet Snippets and Jamie Morgan from Hound Dog Photography

WWW.HUBBLEANDHATTIE.COM

First published in September 2018 by Veloce Publishing Limited, Veloce House, Parkway Farm Business Park, Middle Farm Way, Poundbury, Dorchester, Dorset, DT1 3AR, England. Fax 01305 250479/e-mail info@hubbleandhattie.com/web www.hubbleandhattie.com. ISBN: 978-1-787112-91-9 UPC: 6-36847-01291-5. © Catherine Pickles & Veloce Publishing Ltd 2018.

CONTENTS

FOREWORD

Worzel is the canine ringleader of what can only be described as a modern British, literary reality show circus. Worzel, the rescued Lurcher, narrates, oversees, directs and stars in his own life story, telling it exactly how he sees it. He provides a loveable, sincere, and quite naughty reminder that we never really know what goes on behind your average garden gate.

You can't be sure who you are rooting for in Worzel's tales of life in The House of Pickles (yes, seriously, he lives with the Pickles family – it's like they were already waiting for a canine comedy writer). Mum is always at the end of her tether. Apparently, she's tried her hand at amateur dramatics, and that shines through whenever there is a minor, or major event in Worzel's quite bonkers life. Poor Dad both adores and suffers the trials and tribulations of Worzel. And then, there are 'the cats.'

I spent most of my adult life working with dogs. Much of that time was with a very large dog charity and, in that capacity, I met many Lurchers. Lurchers aren't so much a breed of dog as a phenomenon. Officially, they are a mix between a sighthound, such as a Greyhound or Saluki, and a working dog, such as a Collie, but there are many variations on that theme. The thing is, I don't believe any of it. I mean, I'm sure that Lurchers do come from a 'mum' and 'dad' of some sort or variety, but there's a dash of charisma thrown in, from some special place, that makes them some of the most truly unique characters in the dog world.

Voodoo, Colette, Midas and Titan

Fortunately for me, being honoured with the task of writing this foreword, I am able to share that I was once the furless companion of a very fine Lurcher myself. His name was Ginger Dog. Not a particularly original name, I accept, but certainly an accurate observation. Like so many of his kind, I found him in a rescue centre, where I wasn't looking for a Lurcher at all. He was, allegedly, a cross between a Greyhound and a Border Collie; I'm still convinced that he was half cartoon character and half stuffed animal. Having lived with his Lurcher ways, I can definitely relate to the adventures and mishaps in Worzel's family life. Also fortunately for me, Ginger Dog didn't expose our home life. That story may only be told after the statute of limitations has passed ...

Like many other unique characters in the world, many Lurchers have a colourful and sometimes tragic life. They are over-represented in animal shelters and often misunderstood. That's where Worzel comes in. He is a Lurcher ambassador of sorts. He is fuzzy and fantastic, and through his eyes and words, hilariously translated by Catherine Pickles, he helps the world to know and love a Lurcher. He also manages to define and exemplify doggy diplomacy by walking that delicate line between embarrassing his family and filling them with pride.

Thank you, Worzel, for bringing so much joy to the world, and for shining a much-deserved spotlight on this often overlooked, special type of dog – the Lurcher.

Colette Kase
Head of Animal Welfare and Behaviour for Dogs Trust (retired), and now curator of cool – Worzel you have been officially curated!

INTRODUCTION

Dave and I have been following and sharing Worzel's blog since the beginning, often laughing out loud at his antics on Facebook, or in one of his previous three books. When Worzel gets in a pickle or has problems (which is quite often), Dave will post a funny photo of himself on Facebook, with some friendly and practical 'hadvices.'

I have always wanted a Greyhound called Dave, and before Dave came into our lives, I would come in to the house and shout "Hello Dave." Our manifestation technique worked, and in 2013 he came to live with us when his original re-homer could no longer look after him. It was love at first sight(hound).

Dave is quiet; he only barks when you do something really stupid like leave the gas on or forget to shut the front door. But, unlike Worzel, Dave didn't suffer some unknown abuse or trauma, although the racing industry did give him a fear of loud noises, and cause him injury so that he had to retire.

We can really relate to Worzel's insights of everyday family life. There are many things that all families and sighthounds have in common. It's reassuring to know that you are not alone in the world or that what you think, feel and experience is not 'quite very actual' weird but is 'quite very actual' normal.

Dave is a good-looking boy. Honestly, if he were a human I would consider marriage.

He is extremely muscular, barrel-chested, tall and fit, with doe eyes: he has tattoos.

What's not to love? He is also well-mannered, intelligent, loving, and a great listener, who understands everything I say to him. He is also very polite. After he's eaten he will find me to say thank you, encourage me to rub his neck, and then, once he's burped in my face, he rolls over and goes to sleep. Perfect husband material! All he needs is a job and a bank account!

Just like Worzel it took

Dave

Dave a while to understand stuff. Dave realised after two years that he could 'very actual' climb the stairs. He did go through a stage of sleeping on the bed with us, but we had to put a stop to it as, one Christmas, he slept on my leg and dislocated my knee. I spent a week walking like a peg-leg pirate!

I would actively encourage anyone and everyone to rehome a sighthound. They will bring you so much joy, they will make you giggle with their daftness, like when they tap dance on the wall with their feet while they are sleeping; or with their random roaching and demands for belly rubs. They will absolutely steal your hearts (and sometimes your shoes). I can honestly say that my Greyhound, Dave, and Worzel are the best types of dog in the whole wide world, and I would rescue more, big, lolloppy fur babies if I had room, and more money for cheese.

Equally, I would encourage everyone to fall in love with the Worzel series, follow him on facebook, and enjoy every second that you spend in Worzel's world.

Catherine Eyre
Holistic Therapist and Sighthound Mummy

Catherine and Worzel

Catherine Pickles is a full-time family carer, writer and blogger. Her blog about Worzel reached the final of the UK Blog Awards in 2015, and she is also a finalist in the 2018 Animal Star Awards.

Worzel Wooface is a Hounds First Sighthound Rescue dog who likes walking, spending time with his family, and chasing crows when given the opportunity. His current hobby is chewing wellies on unmade beds. He lives in Suffolk.

Catherine has fostered numerous sighthounds for Hounds First Sighthound rescue. Her hobbies include sailing, walking, gardening and amateur dramatics, most of which she likes to do with Worzel (apart from gardening, but she doesn't have much choice about it, and the amateur dramatics, which he would hate). She, too, lives in Suffolk, with her husband, Mike.

Dedication

For Niall Lester, Animal Warden extraordinaire and founder of New Hope Animal Rescue

January

January 1

It's Noo Near's Day. After last year, I has decided that trusting Noo Near's Day is a Bad Fing, and I shall be very happroaching this one With Caution. Last year, I woke up and discovered that during the night somehow I had managed to try to chop my blinking head off. If that had happened to you, I don't fink you'd trust the next Noo Near's Day neither.

Currently, as far as I can tell, everyfing on my boddedy is hexactly where I did leave it, but I has not have dunned testing this out completely yet. My current position – hupside down on the bed, with one leg helegantly stretching up in the air and another one delicately shoved up Dad's nose – is not wot you'd call mobile, so it isn't hexactly a fair test.

How-very-ever, I has got no plans to make any sudden moves: I are very comfortababble on this big bed and if somefing is going to hunexpectedly fall off my boddedy, it can do it later after I has had a bit more sleep ...

January 1 (ten minutes later)

I has just binned rudely and quite hunnecessarily hejected from the bed. Mum says I are a great-lump-who's-been-lying-on-her-legs-all-night. And could I get out of the way cos she needs a wee. So, quite very earlier than I planned, I are able to report that everyfing seems to be working when I are the right way up. Heven if I don't want to be. In betterer noos, I has dunned my morning shake and my head didn't fall off.

Now that I has established that I are alive and well, it is my himportant work to Get Mum Up. Currently, she is not up. She is sitting on the chair in the room with all the water wot I do not like, trying very hard to stay asleep. And finking about getting back into bed to do some more snoozing. And dreaming about David Tennant. But this would be a Bad Hidea, and not get Dad his cuppatea.

If I was a Labrador, gettering Mum up would be easy. I could do some fusey-tastic tail-smack-and-bashing against her shins, and then find lots of socks and knickers to shove into her hands. And down the loo. But I are not a Labrador, I are a Lurcher and, most himportantly, I does not like the room with the chair and all the water, so there is NO WAY I are going in there.

Instead, my cunning plan to Get. Mum. Up. is to leap back into the bed and find the dent she made whilst she was asleep and fill it up. With Me. And then wriggle around until my head is on her pillow so that I does look as sleepy and comfy as I can. Then, when Mum staggers back into the bedroom with her head full of going-back-to-sleep-and-David-Tennant, she is met by a vision of dorableness that is Worzel Wooface. And Dad as well to be very actual fair. Mum says Dad is dorable too, but honly when he isn't snoring.

Usually, when Mum sees how comfy I are, she does decide that disturbing me would be quite actual cruel and artless. So then she sighs and mutters that she ought-to-get-up-anyway, and leaves me and Dad in peace.

January 2

Not gettering up first in our house is hessential. I tried gettering up first once and it was Orrendous and a never-to-be-repeated hevent. For a start, hespecially at this time of year, it is flipping freezing downstairs, but mainly, there is the cats.

In very general, if I do concentrating very quite hard, I can tolly-rate cats. But honly when I are fully awake and on High Lert. And they has binned fed. A sleepy boykin and hungry cats is a quite actual bad comby-nation, and there is Five. Of. Them.

Gipsy, the senior cat, generally behaves with Digger-nitty in the morning. She sits like a Hemporess, scowling down at everyone, and generally creating a Hatmos-Fear. She doesn't need to do Complaints to the Management about the fact that there is No. Food. because Mouse does it all for her, like a bonkers secretary, flapping and having hystericals. As soon as anyboddedy appears in the morning, Mouse starts wailing. And running backwards and forwards in front of the kettle, to make sure that no-one can touch it before food has binned got out. Any attempts to fill the kettle before there is food in the bowls will be met by bashes with her claws. And more wailing. And a lot and a lot of all-right-Mouse-will-you-give-it-an-OW-you-COW-rest. It's simpler just to feed her and forgot about time and motion studies. It doesn't hurt as much.

The cat food is in the larder and the larder door creaks. In very actual fact it creee-AYAEEEK-s. So loud that even Frank, sleeping upstairs in the room with the chair and all the water, can hear it. Frank doesn't often run but he does first fing in the morning, and it isn't somefing you want to get tangled up with. Frank is fuge. And not that clever to be honest so he can really honly fink of one fing at a time. So when Frank is finking of food, he isn't finking of the fings he might bump into, or even lookering where he is going. And anyfing he bashes into is going to come off worse. Much worse. So it's best just to stay out of his actual way. He's like a runaway train, honly ginger.

Depending on wot happenened after dark, Gandhi will either slide in through the cat flap when he hears all the creaking larder door noise, or he'll be passed out in the shed sleepering off last night's hexcesses. Gandhi is my bestest cat and a Onorary Lurcher. He was borned in this house and so I did have some hinfluence in his formy-tive years. But he's also a murdering bunny-bashing killing machine, and quite very often he is far too full of last night's rabbit to be needing cat food.

Mum's hattitude towards Gandhi changes on a halmost daily basis and centres around the centre bit of a rabbit. I doesn't know wot it is called but there is a bit in the middle of a rabbit that doesn't taste nice. I have had a good sniff of it and it is yuck. I fink it was hinvented so that cats can do showing off to each other how good they is at hunting, cos Gandhi always leaves the showing-off-bit right where the other cats can't fail to see it. Hunfortunately, Mum very quite hoften fails to see it and ... sliding on a squidgy hinside of a bit of rabbit first fing in the morning with bare feet isn't somefing hoomans like to do. So Gandhi is either Mum's fluffy little baby or a lot and a lot of other words beginning with B.

Mabel doesn't come in for her breakfast. She sits on the fence until Mum realises she's missing. She isn't missing a-very-tall, she's waiting patiently for Mum to remember her so she can be carried into the kitchen and presented with a bowl of food away from the other cats, and nowhere near the kettle and not on top of the microwave. Three inches to the left of the vent for the fridge is her ideal spot so that occasional gentle wafts of warm air can tickle her tummy whilst she eats. And if even one of these fings is not arranged to her quite very actual satisfaction, she pretends she can't retract her claws and hangs off Mum's dressing gown until everyfing is Just. So. Mum finks Mabel is sensy-tive and timid: the rest of us fink Mabel is taking Mum for a ride. Literally.

January 3

There used to be two nearly growed-up hoomans who lived in my house. Mum says there is still one but I are finding it very hard to believe. I hasn't seen the previously ginger one since she camed stumbling through the back door on Noo Near's Day. Wearing stilts. She was a bit quite actual wobbly on her stilts, and was muttering about my-feet-hurt and needing to cancel her cashpoint card. Then she crawled up the stairs and I has not seen her since.

I have founded her stilts, though. They is black and about six inches tall. Currently, one of them is hidded in my bed and the other one is in the living room (I are saving it for later). I are halmost sure it is my dooty to make these stilts not wearababble, but the problem is they is quite very solid and also covered in beer and sand, neither of which is tasty. But from the moaning about my-feet-hurt I do fink she would be glad if they wasn't around to tempt her to wear them again any actual more. You can never quite tell with the previously ginger one: one minute she finks somefing is actual terribibble and the next minute it is very quite wonderful. So, to be on the safe side, I has decided to have a bit of a nibble of one of the stilts and leave the other one actual alone until she does making up her mind. Or comes out of her bedroom. Or proves she is still alive. One of them fings.

January 4

The previously ginger one IS alive. And awake. Hand yelling. Happarently, chewing one shoe does not leave any room for making up her mind; if you chew one shoe you might as well chew them both. I did not know this fing. So now they has both gonned in the bin and there have been lots of words about Mum-you-need-to-buy-me-a-pair-of-shoes and did-you-eat-my-cashpoint-card-as-well wot I do fink is most hunfair. She dropped that in Norwich back when it was still 2017 and she didn't have a Ned Ache.

****FORTS ON SHOES****

🐾 Shoes are complercated fings wot hoomans stick on their feet and then spend most of the time complaining about

🐾 Dogs do not generally need shoes, mainly cos they never wear them so the bottoms of their feets do turn into shoes. This is somefing hoomans might want to fink about a bit more. Then they wouldn't need to go to shoe shops and there would be more monies for treats

- The honly shoes that hoomans wear that are himportant are wellies. The rest are either pointless, tasty, painful or involve peoples going out without you
- Apart from slippers. Most hoomans wear slippers hindoors and that means they will be spending the day hindoors with you
- If a hooman is going to wear shoes, in very general they should match each other. Unless someone has lefted a gate open and then anyfing will do. Even nuffink. Even if they have to cross a gravel path
- There is a special song that peoples sing when they has got to cross a gravel path with no shoes on
- Shoes come in lefts and rights. Two lefts do not make everyfing alright
- Most hoomans have two feet. Most girl hoomans seem to have 76 fousand shoes. None of which are comfy. Or match their dresses
- The honly bit of clothes that Dad truly owns wot can't be stolen by Mum is his shoes. Unless she needs to go to the freezer in the shed
- Dad keeps all his shoes underneath Uff the Confuser so he doesn't lose them. Mum says Dad should keep them hupstairs in the bedroom like she does but then she wouldn't have anyfing to wear to go to the freezer
- The soft hinside of a shoe is a food. I has got no hidea how it does grow or how it gets in there but it is fabumazing

January 6

The fuge ginger boyman is not at our house at the moment. I would know if he was cos I would be able to smell him. And hear him talking wherever he was in the house, and from the bottom of the garden, and also be tripping over the stuff he has lefted everywhere. But fings don't look like the Hapocalips and that is because he has gonned back to Universally early, heven though lessons have not started yet. According to the previously ginger one, he's gotta girlfriend, and Mum won't let him bring her home to say hello luffly boykin to me and the rest of the famberly yet, which is quite actual rude I do fink.

Mum says it isn't rude; it is to stop fings gettering complercated and hawkward. Having childrens is hard a-very-nuff but then they start growing up and having relationships. It's the bit of being a parent Mum never actual fort about, and it is Not. Easy.

Sometimes, the fuge ginger boyman does pick nice girlfriends, like the fuge ginger boyman's first girlfriend, Fizzy. We all liked Fizzy, hespecially me, but then the fuge ginger boyman did growing up and so did Fizzy, and as well as growing up they did both do growing apart. All this growing stuff sounds like a bonkers plant in the garden to be actual quite honest. And then they did splitting up wot sounds Orrendous. And painful.

Wot it was. Everybodddy in our famberly got very actual fond of Fizzy, and then we had to get used to her not being around anymore. And not making cuppateas wot she was quite actual exerlent at.

After Fizzy there was The. Other. One. And that was Orrendous. I did not like The. Other. One. I really, wheely fink peoples should do more paying hattention to me sometimes ... I coulda tolded them that The. Other. One was going to be a disaster. She made me nervous and hanxious, and I did a lot and a lot of hiding from her. And in the end she was horrid about the fuge ginger boyman's friends, and heven more horribibble to the fuge ginger boyman. And her eyes were too close together. I didn't notice that bit to be very actual honest, but Dad did. None of us fort Dad was really paying attention but it turns out he can sometimes, very actual hoccasionally find somefing Not. Nice to say about people if he tries quite actual hard and that person has hupset His Boy. It got so very bad that the fuge ginger boyman did fink he might not be able to carry on at Universally. That Bad.

Fortunately, the peoples he was living with did a lot and a lot of helping out, and sending Mum secret messages, so heven though they is all history students they did manage to not be a shambles about this fing.

So Mum has said that she doesn't want to meet the fuge ginger boyman's noo girlfriend just yet. Cos if she is nice and they do stop being acupple then she will be hupset, and if her eyes are too close together then Mum isn't sure she will be able to not say somefing. Not after last time.

And if they is still together after six hole months, and the history students haven't sended Mum any halarming messages about this one being a psycho-witch-from-hell, *then* she can come and visit. But not until then.

January 7

Wot is the actual weather like where you is? Mum and me did decide we HAD to go out today, even though fings looked a bit gloomy and chilly. But by the time we did get to the beach, it wasn't any of those fings. There was no wind wot I hear other people in the North of Ingerland are having to deal with big time badly. And the sunshine was nearly warm. I does hope everyboddedy is keeping safe if they has bad weather, but here everyfing is just a-very-bout Costa Del Suffolk.

January 8

Dad says I are not allowed to do showing off about our weather. Happarently, I has tork-tit-tup and now it is my fault that we has binned promised cold-wet-and-miserababble, and maybe heven snow by the end of the week.

I are nearly quite very certain that although I are a himportant and luffly boykin in my famberly, my hopinions about the weather aren't hexactly glow-ball noos. And I do fink it is actually the fault of Mark-the-man-on-the-radio, who tells everyboddedy everyfing and it is him who is the tork-tit-tupper.

January 10

Worzel weather hupdate: it's blinking freezing. We've had a bitta snow but not a lot. There is a nasty cold wind and I are staying in bed.

According to the tork-tit-tupper, it isn't the snow we should be worried about. It's the wind and the snow melting and the tides.

It is a quite actual well-known fact that I are a Nable Seaman so I do know tides hexist, but I hasn't got a clue how they do work. Dad says it is somefing to do with the moon. I find this actual quite hard to believe. I fort the moon was there to help hoomans see in the dark. And not do tripping over the curb and smacking their faces on the concrete. And then frightening me big-time-badly with their hystericals. But happarently, as well as stopping hoomans making hidiots of themselves when the neighbours are watching, the moon does also drag water around the Planet. And Hunfortunately, the moon is going to drag all the water past Southwold tonight with a bit of help from the wind, and then there will be too much of it to stay in the sea and it's all going to end up in the boat yard.

At least that's wot Mark-the-man-on-the-radio says. And he says he got it from the poxy guv'ment. So it must be true, mustn't it?

Dad's not convinced. He may not be Mark-the-man-on-the-radio or the poxy guv'ment but he has hexperience, and also bits of paper and a confuser called Uff wot can do sums ...

January 11

... and my Mum who sends messages to Mark-the-man-on-the-radio to tell him that Dad says the poxy guv'ment is overreacting big-time-badly. So now, as well as Dad trying to stop the water going into the boat yard, he's got to do talking on the radio about it at the same time!

Mum finks it's marvellous. Mum will know hexactly where Dad is and that he is still alive, and also when he is going to come home. And when she needs to put the kettle on.

January 12

Dad was fabumazing on the radio. He did do three reports from the boat yard about the water and the flooding, HAND he managed to do Complaints to the Management about how dangerous it is when the poxy guv'ment torks-tit-tup big-time-badly. Cos the water did come into the boat yard a little bit, but it came up to hexactly where Dad's bits of paper and his sums said it would and not nearly as far as the poxy guv'ment was wailing about. According to them, we was going to need a Nark. I aren't sure wot a Nark is but it is all very okay cos we didn't need one. But next time, peoples might fink the poxy guv'ment is hexaggerating again and not do anyfing, saying a lot and a lot of stuff about a crying wolf.

Wot the Heckington Stanley has a wolf got to do with floods? Is it somefing to do with Narks? And why is he crying?

January 13

Ohhh ... a crying wolf is all about saying somefing has happened when it hasn't, and then getting upset when the fing *does* happen and noboddedy comes to help. Mabel does that but she is not a wolf. She is an ickle cat who finks that everyboddedy is out to do her damages. Currently, her hassassin is Frank, and every time he does come near her she hisses and spits and crouches down. No wonder he wants to biff her ... Heventually, Frank will get actual fed up with all this he-is-trying-to-murder-me hissing and fluffing, and will really bonk her and noboddedy will believe Mabel. Her current hexclusion zone is about five feet which, in a quite very small kitchen, doesn't give an actual lot of room for man-hoover.

All Frank wants to do is get a drink of water from the bowl in the corner of the windowsill, and Mabel is making it himpossibibble.

January 14

All of the cats, apart from Mabel, has decided that my drinking bowl in the hallway is a betterer hoption than the bowl on the windowsill. Now, my water bowl is halmost hempty and Mum. Has. Not. Noticed. Having an hempty water bowl is not somefing I has dunned hexperiencing before. In very general, I does not do drinking a lot and a lot of water: I have a raw dinner so I don't get firsty in the same way as doggies wot have to eat brown-balls-full-of-wheat-and-rubbish. Usually wot happens is that Mum notices that the one on the windowsill is hempty and then checks mine after she has filled up the windowsill one. But that one is actual quite full and I has got no hidea how to let Mum know that fings have dunned changing, and my honly hoption is to rely on Frank gettering fed up with Mabel and biffing her, and then Mum putting two and another two together and realising that there is a chain wee-action going on. And not a lot of weeing from Worzel Wooface. And finally fillering up my water bowl. I aren't holding out much hope ...

January 15

My water bowl is now completely hempty and I starting to wonder how I can get onto the windowsill in the kitchen. I does know that in feery it is possibibble. Harry, my last foster brother, did manage to do this fing very quite hoften. He did just jump up there and pinch all of the cat food whenever he did get the chance. But the fing is, I aren't wot you'd call a Nagile dog. Being Nagile isn't one of my talents. And also the windowsill is skiddy and shiny and still has a Mabel satted on it. Fortunately, my breakfast this morning was quite juicy and so I are not dying of first – yet.

January 16

Last night, I wented round to play with Kite, but before I did doing any playing or running around, I did go and find Kite's water bowl and had some ignormous slurps of her water. And as Kite honly has to do puttering up with one cat, she does not find herself being coll-actual damage in the hole water bowl/windowsill/hassassin hystericals-dramaticals wot lead to my water bowl being hempty. And I are quite actual pleased to say that Mum did notice all my slurping of the water bowl and did become very actual concerned about Worzel's-drinking-a-lot.

When we gotted home, Mum did lookering in my water bowl and sawed that it was hempty, and himmediately filled it up. And there were lots of words about sorry-Worzel-Wooface and why-are-you-so-firsty? So then I did get some more to drink. And just when fings were lookering very quite hopeful that Mum would do realising that Mabel had dunned taking the water bowl on the windowsill hostage, and that after I had had my slurps of water there was a queue of cats waiting their actual turn, she did hurry off into the Hoffice and began lookering up Dog Drinking A Lot of Water on Uff the confuser.

Now, I can tell you for blinking nuffink, Uff the confuser was never, hever going to come up with Mabel finks Frank is trying to kill her so she's hissing every time he goes onto the windowsill to get a drink of water, so none of the other cats will go on the windowsill neither in case they get turned into coll-actual damage, so they is all drinking out of my water bowl, which is why there's no water in it for Worzel Wooface. Cos Mr Google and Uff the confuser might be quite actual clever, but they'd have to be flipping side-kick to work out that lot. Or actual be here. In the kitchen. Watching wot is going on and not asking Uff the Confuser the wrong question. And then not actual surprisingly, getting the wrong answer.

Mum is now convinced I have got Die-A-Bee-Tees.

January 17

Today, Mum has trotted around after me waggling a pot every time I has gone into the garden. Her cunning plan is to get a wee sample from me so that she can take it to Sally-the-Vet to see if there is sugar in my wee.

If anyboddedy finks I are going to do weeing in a pot they has got another fink coming. I are quite private about my weeing hactivities, and I are not weeing within arm's length of anyboddedy. And after this morning's not-too-subtle-creep-up-on-Worzel-and-catch-him-when-he-is-not-lookering attempts, I aren't going into the garden any more, and I aren't going within ten arms' lengths of anyboddedy. I are staying in bed and hiding.

January 19

Mum's decided I can't have Die-A-Bee-Tees. Not 'can't' as in there-is-hevidence-I-does-not-have-it but 'can't' as in Mum is never going to get the hevidence so I are not actual allowed to have it.

Honly Mum, Dad reckons, could fink that there is any logic at all in this, and I was quite very hopeful that Dad would stop talking and would p'raps do noticing that I are currently sharing my water bowl with all the cats (apart from Mabel), and that the main reason I 'can't' have Die-A-Bee-Tees is because I haven't actual flipping got it. But he was far more hinterested in laughing at Mum's logic and questioning her hintelligence.

It has now becomed Dad's job to collect the wee sample if he's-so-somefing-beginning-with-B-clever.

January 20

Hoccasionally, the previously ginger one is quite very actual useful, hespecially when she is going through one of her 'I can't sleep' phases. Last night, when I hearded rummaging around in the fridge I went downstairs to hinvestigate, and I had an exerlent plan about barking at the hunexpected visitors. But the smell of frying bacon made me realise that there was no need to do that fing: in actual very general robber-dobbers do not use frying pans for cooking. So instead I did asking to go outside for a wee in the dark On. My. Own. with no pot waggling.

Whilst I was out there, Frank and Mabel started up their windowsill wars again, and Mabel somehow managed to end up quite actual soggy.

Mabel has the sort of fur that soaks up water really, wheely well, and I do fink the previously ginger one felted actual quite sorry for her, backed up into the corner, and then backed up into the water bowl until she was halmost sitting in it.

After the previously ginger one had fished Mabel out of the water bowl and tried to dry her off a bit and got scratched and hissed at for her troubles, she did refill the water bowl. But instead of filling it up with a jug, the previously ginger one lifted it off the windowsill, emptied out the bits of Mabel wot were in it, and then put the water bowl back on the windowsill but not tucked up in the corner. Instead, she has putted it right in the middle of the windowsill. Now all the cats can get to the water bowl, Dad won't have to chase me round the garden with a Waggling Pot and end up feeling as stoopid and redickerless as Mum, and none of us animals is going to die of first or get haccused of having Die-A-Bee-Tees.

The previously ginger one has No Hidea wot she did just do. And she still can't get to sleep.

January 21
****FUGE NOOS****
Mum and Dad are going to Marry Kesh to celly-brate being married for ten hole years. And not killing each other. Although I has got to say that when Dad did do suggesting this and did the choosing of the place to go, Mum did nearly pass out and need to sit down for a moment.

Fing is ... Dad doesn't do romantic jesters. Ever. I don't fink it's got anyfing to do with Mum: his brain doesn't work that way. Like I can't do heelwork to music. Mum and Dad's hanniversary isn't until July, but July is no good cos that's in the summer when sailing happens, so instead they is going in a few days whilst it is cold in Ingerland and hopefully a lot warmer in Marry Kesh.

Mum says I'm not allowed to come to Marry Kesh, mainly cos I would not like it. And that she and Dad will be flying. I are very quite surprised by this cos as far as I are aware neither has ever showed any hindication that they can get off the ground for more than about a second. And honly then when Leeds scores or Mum can't reach a high shelf. You would fink if they could do flying they'd use that talent a bit more quite very actual often ...

January 22
Dad says I have got it all wrong and they will be gettering on a plane wot is a machine that lifts off the ground, and someboddedy else will do the steering. I find this heven more hard to believe cos Mum is a Control Freek, so I can't hunderstand how she is going to sit in a metal box and let someboddedy else steer her for four hours at a billion miles an hour to a strange place. Through the air. When Dad drives the car, she secretly puts on Mr Sat Nav, heven though Dad says he knows where he's going, just to make sure. And so she can offer hirritatingly helpful suggestions about it'd-be-quicker-if-you-went-this-way, and I-know-where-we-can-park. Mum must really, wheely want to go to Marry Kesh.

Whilst Mum and Dad are in Marry Kesh, I will be staying with Gran-the-

Dog-Hexpert. And Iona. And all her other doggies but they is mainly quite actual okay. Baillie is a bit too actual sertive for my liking, but the others are all either playful or hancient. But Iona is a fishwife. Gran-the-Dog-Hexpert says I are marvellous for Iona. Most of the time, Iona is bottom-of-the-heap and gets bossed about by all the other doggies and has to wait until last to go through a door and all that sort of fing. But when I come to stay, Iona has someboddedy to boss around: me. And boy, does she take full hadvantage of the hopper-tunity.

January 23
Mum keeps saying "But man, proud man, Dress'd in a little brief authority" like it is supposed to Mean Somefing. It's from a play called *Measure for Measure* and Mum did it for her Oh levels. It's the honly bit of her Oh Level she can remember, so she's very actual glad it's come in useful after thirty years. It's a posh play by a man called Shake-a-Spear, happarently, and there's a man in it who gets to be in charge for a little bit. And he messes it all up. Big-time-badly. And then the King comes back and saves the day and ... Mum can't remember if the man that messed it up big-time-badly gets deaded or not. I fink Mum is getting completely hover-hexcited about going to Marry Kesh and she has forgotted that I are a dog and ...

- Iona isn't a man, she's a tiny ickle Cavalier King Charles Spaniel
- Hiambic Pentameters be actual quite lost on me and Iona. In fact, on all of the doggies
- I don't want Iona to do getting deaded; I just want her to Stop. Barking. At. Me!

January 26
Mum and Dad are back from Marry Kesh. They did have a fabumazing time, and heven though it was January, it was warm a-very-nuff to walk about without a coat. They did so much walking and hinvestigating around Marry Kesh that Mum's step-counting fing on her phone started making pictures of fireforks.

Before they went, Mum was a bit actual concerned that she might see hanimals in troubles or not being looked after properly. She did not see this a-very-tall, she are glad to say, but she did see fundreds of cats and all of them did seem to have all their gentleman bits still attached to their boddedy, which did probabably very hexplain why there was so many of them about.

Mum did refuse Point. Actual. Blank. to go into the main square where there were people being hidiots with snakes because she did Not. Happrove. The hole hexperience was like visiting another world in many ways, but they very quite soon got used to the different world and managed to not get lost or frightened. Both Mum and Dad are very actual pleased to be back in Ingerland to see their luffly boykin. But mainly, they would like a proper cuppatea ...

January 27
Dad says he's dot-a-dold. He keeps doing sneezing and spluttering but he has gonned to work cos he says he's going to feel rubbish wherever he is. And he

says he's dot-a-dob he can do on his own without hinfecting everyboddedy else. I don't fink dogs can catch hooman dolds. There is lots of nasty fings dogs can get wot make them feel hunwell, but I has not never had a dold, even though all the hooman's here have. Mum says she doesn't want to catch the bug that Dad's actual got, so I are In Charge of giving him kisses and cuddles until he is betterer. Maybe this is why dogs don't get dolds. So they can do this fing ...

January 28

I met two Samoy-Teds at the beach today. I has never metted a Samoy-Ted before, and to be very actual honest, they did look just like Teddy Bears. Fortunately, I didn't do getting them muddled up with teddies cos my track record with soft toys is a bit kinda, well, terminal. After a bit of hello-luffly-boykin sniffs and hunt-the-gentleman-bit under all the fur to make sure they was actual doggies, we did have a fabumazing time making friends and playing in the puddles that the sea made on the beach.

I should also do confessing Mum's sins. Mum says this is not actual necessary, but as this is my diary and my forts and feelings, I shall do that fing.

POO BAGS

Mum forgot the poo bags
So when I did a poo
She had a blinking panic
It was a right to-do

She panted up the sand
And got one from a friend
But then she couldn't find the poo
When would the drama end?

The moral of this story
(If it needs an end like that)
Is always bring your poo bags
Or you'll look a blinking prat

January 29

I did meet some super noo friends today. Bertie was with his famberly, and they had comed all the way from Middlesborough to see me! Well, nearly ... they had comed to visit their famberly in Lowestoft, and then they did pop down to meet me in Southwold. Before we lefted the house, Mum did look halmost respectababble, but then she put on her noo birfday coat from Marry Kesh and her noo birfday wellies from Dad ... our noo friends musta fort she was a part-time clown.

Bertie was a super friendly boykin and did exerlent come-in-the-general-right-direction, but he did not want to get his feets wet in the sea wot was actual quite ruff and frilling. There was even some brave chilly surfing peoples in the sea!

For THE QUITE *very* actual LOVE of **Worzel Wooface**

January 31

Gipsy-the-Cat is a Nigma. We is never quite sure where she goes during the day apart from Out. Sometimes when Dad is driving home from working down at the harbour, he does see her hiding in the hedges, and then she'll try and race the car home cos she finks it is teatime. Racing a car home is probababbly not a fabumazing hidea, to be quite actual honest, hespecially for a cat wot's got asthma, but she's been doing it for nearly thirteen years and you can't tell Gipsy actual nuffink. She is In. Charge. and does her own fing. Apart from breathing. She's rubbish at that so she has to go to see Sally-the-Vet about once a month for a hinjection.

Getting Gipsy to the vet is A Mission. It hinvolves an ignormous lump of luck that she will decide to come home the morning of the vet visit, and then a]quite very actual ignormous lot of hoomans remembering to shut doors and the cat flap. And then, once Gipsy is trapped in the house, getting out the cat basket. If Gipsy spots the cat basket before she is stucked in the house, she does bogging off for two days, and Mum has to do hembarrassing phone calls about how she's messed it all up and can she try again tomorrow? Dad is very not allowed to get hinvolved in Mission Gipsy To The Vet cos he says fings like Gipsy-doesn't-want-to-get-in-the-basket, and poor-Gipsy-Mummy's-being-mean-isn't-she, and Mum says it is really, wheely hirritating to hear this when you've got twnety claws himbedded in your tummy. And some of the holes is starting to bleed.

Anyway, Gipsy did heventually go to Stoven Hall Veterinary and Rehabilitation Centre to see Sally-the-Vet, and after last year's worries and panicks about her, she is doing quite very actual well, all fings considered. Betterer than Mum's belly, that's for sure ...

FEBRUARY

february 1

Sometimes, it is my himportant work to edercate peoples about the Ways of the World. And about being a Lurcher and how important sofas are. Stuff like that. But sometimes I do fink it is himportant to tell people about other facts of life that are Not. Nice. If you has had a bad day and there is no wine, you might want to not read my forts and feelings today, and I will not be hoffended a-very-tall if you does skip over it and come back later.

Today is World Galgo Day. Galgos are Spanish sighthounds. They get treated very actual badly, and today is the last day of the hunting season when fundreds and fousands get abandoned. Or worse. Some of the peoples who use the dogs for hunting do hunspeakable fings to their dogs if they is rubbish at hunting, wot is Orrific.

Fing is, the reason I ended up in such a blinking muddle when I was an ickle baby puppy was due to the fact that I did show no talent for hunting. I was too actual scared and worried about the hole hidea of deading fings wot are alive, so I did get passed around all sorts of peoples and places until I ended up heven more scared and even more hanxious. None of that was hideal and fings were very quite actual bad for a while.

Some of the hunting bits I has got talents for. I can see fings and I can chase after fings quite actual fast, but when it comes to the hole deading fing, I has got no skills a-very-tall. The honly fing I ever, hever dead is cushions or teddy bears. And shoes. In very fact, I are an hexpert at deading fings wot were never alive in the first place.

So after I has dunned the seeing and the chasing, I does forget wot I are supposed to be doing and generally end up giving a play bow to whatever hanimal I has just chased halfway to Scotland, and apart from puffing-out the hanimal and possibibbly, probababbly, almost definitely ruining his day, the hanimal does at very least get to have another actual day.

And then Mum, or most very often Dad – cos he is rubbish at the hole lookering-at-the-horizon-and-checking-there-is-nuffink-worth-chasing fing – catches up with me and calls me a plonker and we does go home. And Dad is frilly-sofical about No Harm Done, and walks in through the door wittling on about Worzel being Mum's Dog and I've-lost-another-hat-chasing-your-flipping-dog-through-the-woods.

But he doesn't fink I has bought shame on my famberly for not doing the deading fing, and he doesn't believe that I are unlucky like Galgo owners do sometimes fink. So he doesn't do murdering me by hanging me from a tree.

If I had binned a Spanish Galgo wot is very quite similar to an Ingerlish Lurcher that might have binned wot happened to me. And although it isn't very actual easy for me to tell you about these fings, I fort you should know so that you can do wot you can to save the Galgo from this hawful fing.

For THE QUITE ^{very} actual LOVE of Worzel Wooface

februarч 4

Dear Mum

I does very actual happreciate that we has a bond and you is the most himportant person in my life. Wot's mine is yours and wot's yours is quite very actual mine. Hespecially the bed and the chicken wings.

But there is limits to how far I does want to do bonding with you. And it do stop way before we does get anywhere near my toes, my hiballs, and hinside my ears. You is not allowed to do finking that we is bonded in those areas. We is not. You can hinspect those areas from a distance of my choosering, like across the room. Or from the top of the stairs. And if from that safe distance you does still have concerns, please do taking me to visit Sally-the-Vet, who spended seven years learning how to do hinspecting with her hands as well as her hiballs.

From your luffly boykin

Worzel Wooface

februarч 6

Dear Mum

And we is definitely Not. Bonded. in the gentleman bit department. And hespecially not the tufty fluff wot grows out of the end of it. Even if it is all tangled up and stucked together and a bit actual smelly. So please do putting down the scissors.

From your luffly boykin

Worzel Wooface

februarч 7

Dear Dad

Fank oo for volunteering to be trimmed instead of Worzel Wooface. Well, not in the same place hexactly, and probababbly not actual volunteering but not moving as fast as me. As far as I are concerned, that is the same as volunteering and until Mum loses the scissors it is every man and luffly boykin for himself.

Sorry about that fing ...

From your actual quickerer and very hiding at the bottom of the garden luffly boykin

Worzel Wooface

februarч 8

Dear previously ginger one

Fank oo for actual hexplaining wot manscaping is all about. Dad says he would like to un-know this fing. And he is never, hever letting Mum do it to him, no matter how many drinks of Cider she pours down him. I would like everyboddedy to know that I are a dog and not a man so I do not qually-fry for manscaping neither.

Anyway, me and Dad has dunned a deal. Please will you tell Mum that we will stop hiding in the shed and refusering to come out if Mum Puts. Down. The. Scissors.

That's the hessential bit. Nuffink is going to happen until Mum hands over her actual weapon.

Once Mum is hunarmed and no longer a leaf-full weapon, Dad says he will do somefing about his own hairy Hearholes, and I has hagreed that he can also do trimming my yucky bit of tufty fluff cos his hands don't shake.

And also he is more hexperienced in the gentleman bit department.

From your luffly boykin

Worzel Wooface

22

Pee-Ess: Dad says in future could you keep your maggy-zines away from Mum cos they does contain far too much hinformation ...

February 9

Yesterday Mum had a furry grey hat. Today, she has got lots and lots of furry grey scraps of fluff and a sore throat from screaming, and an Art wot is never going to be the same again.

Cos I deaded it. Not Mum's art – although by the look of her it was a close-runned fing – the grey hat. And happarently, dead furry grey hats wot has binned shredded all over the hall floor look hexactly like torn-to-bits Mouses or Mabels.

Mum did love that hat. But not as much as she does luff Mouse and Mabel, so she has dunned deciding to forgive me and I aren't in trouble for hexploding the hat cos it's just-a-hat-just-a-hat-stop-screaming-you-daft-woman. It is a very quite strange fact that you can get away with actual murder if you hasn't dunned actual real murder ...

February 10

You would fink I would know betterer. Just when I did fink there was nuffink else Mum could find to worry her, she has managed to find somefing else to panic about. According to Dad, if Mum hasn't got anyfing to worry about, she worries that she should be worrying about somefing that she's forgotted. Sometimes, I do fink being hooman and hespecially Mum is Very. Hard. Work.

Tonight, Mum's hanxieties have been all about me and my recall, wot is gettering worse, and not getting any better as the years go by. She says it is happalling and it is gettering worse. It's not like she's stopped making it a Pri-Orry-Tree, but it has gotted to the stage where she is scared to take me out for a walk and let me off my lead for zoomies cos every time she does, I take longerer and longerer to come back.

Dad tried to be helpful but I fink he is wishing he hadn't now because it just meant that Mum did more hand-flappering and squeaking and having hystericals, and there were a lot and a lot of too many blinking words for me to keep up. But the main actual highlights of her hystericals is somefing like this: Worzel could run into a road and get knocked over, or he could get stolen or hinjured and stucked in a ditch and lie there for days and days not able to get out, or could even do dying ...

She's not wot you'd call Oppy-Mistic is my Mum. And now the chance of me going for a walk tomorrow is very actual zero.

February 12

Mum might have a cunning plan. In-very-general, I aren't keen on Mum's cunning plans, but this one is quite actual hopeful. To be quite very honest, anyfing has got to be betterer than the current confined-to-barracks situ-nation wot is happening. She has dunned finding out about Gee-Pee-Ess trackers. Happarently, they is this fing wot can attach to my collar that sends a signal to the Hole Wide World about where I am. It isn't going to give me a Magical Perfick Recall but it will give Mum half a chance of knowing where I am. I aren't sure about this: I aren't sure I want the Hole Wide World knowing where I am. Sometimes, a luffly boykin needs a bit of actual alone time for fortful finking and doing a wee and also bogging off. When I do bogging off in the woods to go and lummox about in the ditches, the hole hidea is that I've Bogged. Off. and I will do coming back when I are covered in black sticky gunky stuff. But Mum says that's tuff; she has a-very-nuff to worry about already, and this will give her some Peace. Of. Mind. The hole being-scared-to-take-me-for-a-walk-in-case-I-get-lost fing is gettering beyond a joke.

So I has got a choice: wear a Gee-Pee-Ess tracker and be allowed some off-lead time and half a chance of sneaking off for some ditch-wallowing, or stay forever on a lead attached to Mum. And Dad reckons it doesn't matter how much she loves me, Mum is never, hever going to go wallowing in muddy ditches with me. And more himportantly, however much he loves both of us, he isn't going to drag Mum out when she gets stuck.

february 13

Last night, down near his Universally, the fuge ginger boyman found a lost senior doggy that didn't seem too very well. At 2am, which is not the bestest time to be a lost dog. Like all exerlent boymen wot don't know wot to do next, he did phone his Mum. My Mum. As you can actual himagine, Mum did have mixed forts about this phone call, mainly around the hole gettering over the fact that noboddedy had died and have-you-been-arrested? Them sorts of mixed forts are quite actual difficult to deal with, so finding out about a lost dog was much, much easier. So, she did tell him to take her home and give her some food. And then she said that first fing in the morning he must try to find out if the dog was microchipped and also hinform the Dog Warden, cos that is The Law.

This morning, the fuge ginger boyman says he is actual quite tired cos the luffly doggy did want to go for a wee at 6am, and then as soon as he did going back to sleep, Mum woked him up to find out how the luffly dog was doing, and had he dunned finding a Dog Warden to tell?

Last night, the fuge ginger boyman kept saying fings like it's-such-a-luffly-doggy and if-noboddedy-wants-her-I-shall-do-keepering-her. But the good news is that about an hour ago the fuge ginger boyman did manage to reunite the doggy with her owners. And now that it is All Over, he's hinformed Mum that she won't be having any grandchildren for Quite. Some. Time. cos the stress and the responsibilititty of lookering after a stray dog for one night has nearly killed him. And he's Not Ready.

In fact, he's not sure now whether he will ever, Hever, be ready. And is that okay with her?

february 14

We has got a tracker and it has arrived! We has choosed one from Huncle Peter at Paw Trax because he does answering hemails quickerly. Mum says it's called custy-more Service. Dad says he is very actual glad the man answers hemails quickly cos he hasn't got a clue about the settings, and that means he hasn't been bombarded with text messages from Mum about wot-does-this-mean and how-many-seconds-are-there-in-20-minutes? Huncle Peter has had to deal with all of that.

But custy-more Service isn't a-very-nuff, happarently. The other himportant fing is whether it actual works and wot would happen if I did get actual lost cos that isn't somefing any of us want to hachieve. The hole hidea is that I don't get lost, and heven though Mum is quite actual himpressed with Huncle Pete's custy-more Service, she still has Trust Issues according to Dad. So tomorrow, Mum is going to send Dad to work attached to the Paw Trax, and Mum is going to pretend to have hys013erricals and tell the confuser hinside it to speed up the number of times it says where Dad is ... to see if she can track him down.

February 15

Dad's taken the tracker to work. I are almost actual certain he has left his brains here, at home, with me. Now, cos I are a dog, I does get the Hole Deal. Mum gives me cheese and it's my himportant work to hang around with her and keep her comp-knee during the day.

But very sometimes, Dad likes to do fings without Mum. Mainly they hinvolve having long, boring talks with peoples about fenders and bow-frusters. wot is all very armless. But there is two fings that Dad is Not. Allowed. to Do.

☹ He is Not. Allowed. to go on motorbikes. Or heven look at them. You might fink this is actual cruel of Mum but it is for his own good, like me having my nails clipped. Dad is Not. Allowed. on motorbikes because he goes too very actual fast on them, and before he did become my Dad he nearly didn't become anyboddedy's Dad cos he felled off one. And the doctors honly just managed to keep his leg attached to his boddedy. You would fink he would have learned his lesson but in-very-general Dads don't learn their lessons, which is why they have to have Mums. Cos Mums are very, very good at learning lessons, hespecially for other people. Honly for other peoples if I are quite actual honest ...

☹ Dad is Not. Allowed. in Rope Shops without Mum. There is no other way of actual saying this but ... My name is Worzel Wooface and I are the dog of a Naddict. Dad is a Rope Naddict. Either that, or he is the Patron Saint of Ropes and String. There isn't a bit of rope that Dad doesn't luff. Even the smallest, scruffiest it-isn't-any-good-throw-it-away bit of rope Dad keeps. And luffs. And ties up into an ickle complercated bundle so he can hang it on a hook and look at it. When we go to Rope Shops, Mum has to do holding onto all the monies cos otherwise Dad just buys more ropes to bring them home to his shed. Even if he has some halmost-the-same bits at home that he's never-going-to-use-put-it-down. Every bit of rope is welcome in our shed

And it's not just the monies, Mum says. We are running out of places to put all the bits of rope that Dad keeps 'finding,' and if she hears the words it-might-come-in-useful one more time I fink I might find I are the dog of a broken home. That's probababbly a hexaggeration but quite possibibbly a broken Dad, or maybe a seriously himmobilised Dad, tied to a chair with some of his blinking-bits-of-string-stop-buying-the-blasted-stuff.

But now Mum has got the tracker. Heventually, the fact that Gee-Pee-Ess trackers will work for Dads as well as dogs will come into Mum's brain, and then his life is never going to be the same again. To be very actual honest, it is going to be like mine. Without the cheese.

February 16

Mum's very actual attempts at being a Master Spy are over. Before they had even begunned if I are actual honest ... She has been beated by teck-nology I do fink. This is more plite than being a daft cow and a dipstick, which is wot Dad called her, but they is both actual true ... Mum tried to be hoffended by the hole daft cow fing but it is hard to argue about. Hespecially as she did forget to charge the tracker fing and sended Dad out with a nice little box with a paw print on it that was about as useful as a necklace.

So now Dad is In. Charge. of Master Spying. And the tracker is getting some helectricity putted in it. Tomorrow I will be able to maybe, possibibbly give you a hupdate on how it all works. Probababbly perfickly once Mum isn't hinvolved ...

THE TRACKER (A NODE TO DAFT COWS)
I got a new tracker
It hangs from my collar
And if I get lost
Her dog Mum can foller ...

At least that's the feery
It's a simple hinvention
Unless Mum's hinvolved
Wot I have to quite mention

My tracker won't track me
Unless it has leck-trick
Dad called Mum some rude words
Like daft cow and dipstick

The tracker found Dad
And showed where he went
It's working quite perfick
Mum's quite actual content

Dad is still outside
In the rain he does roam
Cos Mum hasn't told him
That he can come home

26

He finks she's still testing
The tracker's nifty trick
But she's not yet forgotten
He called her a dipstick

February 17

The previously ginger one does not get lost so she does not need a tracker.
Not when she has a phone that she is permy-nantly attached to. She hardly ever
goes out because of her Men-Tall Elf, so where she is is not really much of a
mystery but still she makes a ding noise. Mostly about cuppateas and can-she-
have-some-bandwidth?

And when she does go out and Mum does sighing with relief that she
has finally dunned leaving the house and is seeing her friends to go dancing,
she carries on dinging about where she is and wot she is doing, and also letting
Mum she hasn't drunked too much Cider.

Fing is, the previously ginger one always forgets wot time it is, so
Mum will be in bed and nearly fast asleep when her phone starts dinging with
messages about how-much-is-it-for-a-taxi and can-Mum-lend-her-some-monies?

Dad finks that Mum should turn off her phone at night so that the
previously ginger one doesn't keep waking her up, but I do not fink that is
going to happen until the previously ginger one is at least forty. Or completely
betterer and not at all actual likely to do having a panic and needering rescuing.

Everyfing is quite actual complercated to be honest: on the one hand,
the previously ginger one is a hadult and needs to learn to do hadult fings like
remember to save a-very-nuff monies for a taxi home. But on the other hand,
she did missing out on Being A Teenager, and a lot and a lot of the normal stuff
that hoomans do when they is nearly growed up but not quite.

So it is hard to get a balance: Mum wants her to be responsibibble and
the previously ginger one probababbly, possibibbly may be able to look after
herself. But if it All Goes Wrong it will go really, wheely wrong, and noboddedy
wants that to happen. In the meantime we have just got to actual put up with
Mum having an art attack every time the phone goes ding in the middle of the
night.

February 19

Today was quite actual special. I did get to finally meet my looky-like Lurcher
pal, Dodger, and his ickle big sister, Scout. Mum fort about putting Scout down
her jumper and bringing her home cos she said Scout was a cutie. But then
Scout's Mum said she did probababbly smell quite bad. Wot was quite hoff-
putting and not cute, and not wot you'd want stucked down your jumper.

We went to the beach and I wored my tracker, wot is actual not too
heavy, and I does not fink I look like a hidiot. According to Mum it was my
himportant work not to dunk the tracker in the sea cos it doesn't like saltwater.
But as I does never go into the water much past my knees, and they is a long
way away from my neck, that wasn't too much of a problem, hespecially today
cos the beach was covered in FUFF.

I hasn't got a clue who did sticking washing-up soap in the sea, but they did making a right flipping, well, FUFF! I could not do working out if it was safe to put my feets in, so I did mainly jump over it. None of the hoomans could remember wot it was and whether it was dangerous for luffly boykins, so they was quite actual glad I did jumping over it, and quite actual relieved that Scout, Dodger's ickle big sister, didn't decide to try to paddle cos she woulda dunned disappearing hunder it all, and probabably never comed out.

When we got home, Mum looked on her Confuser to find out more about the beach Fuff. It's mainly safe though it smells a bit, and sometimes has algae in it, so I has had to have my feets washed in case I do have A Hunexpected Reaction. All this feet washering and clucking and attention did lead me to have a very quite Hexpected Reaction and now I are sulking big-time-badly and pretendering that I does not have a Mum.

February 20

Yesterday, when I was at the beach, I did an ignormous recall With. Witnesses. wot did himpress Mum very actual muchly. It mighta binned more himpressive if Mum had binned all notchy-lent, like it was a normal hoccurance, but instead, she did squealing like she'd won the lottery.

All day, Mum's binned muttering about Dodger's Magical Perfick Recall and Dodger-doesn't-bog-off. Well, Dodger is the odd one, not me. And me not being much actual good at come-in-the-right-direction shouldn't hexactly be a surprise. I are a Lurcher and it's in my jeans. Some dogs are designed to come ... but Lurchers are designed to go. 'Go' is wot we've binned doing for fundreds of years. It's all the hoomans fault, I do fink. They spended so much time making sure we could do a fabumazing go, but they shoulda put more fort into wot would happen after the 'go' bit, and realised that the betterer we got at 'go,' the further away we'd be when it was time to do 'come.'

Dad says it is the Law of Hunintended Consy-quences. Like marrying Mum and then realising it meant fundreds of cats, two children, a lot and a lot of gardening, and not a-very-nuff sailing. And then he went all frilly-sofical and decided to press his buttons on Uff the Confuser and Stop Talking before there was another quite very actual hunitendend – but actual more predictababble consy-quence – no dinner.

February 21

Beach or marsh? Marsh or actual beach? Sometimes, a luffly boykin does have some very actual difficult choices to make, I do fink.

They is both fabumazing. The beach has smells-washed-up-by-the-sea (and sometimes other more hinteresting fings to be quite actual honest), and the marsh is covered in smells-dropped-by-the-birds, wot it is hessential for me to do rolling in. So when Mum does talking about should-we-go-to-the-beach-or-the-marsh, I do find it himpossibibble to choose.

Sometimes, though, we does not have a choice because the farmer puts cows on the marsh field and then we have to go to the beach, like today. And today, Mum had rather we had not gonned to the beach cos I was An Hemabarrassment. A big, fuge, wriggly and quite very fusey-tastic Hembarrassment.

Being hembarrassed is not as hawful as me getting lost of stolen or squashed on the road, and Mum could see me the hole time so she did not have hysterics, but it was quite actual difficult for her. It wasn't for me, though, it was actual easy, and I did have a fabumazing time, but a man with a Poodle on a lead ended up in a right blinking muddle trying to hang on to his doggy and not get tied up in knots. And when he started to walk off the beach, I did finking I might follow along ...

Heventually, Mum caughted up with me and the Poodle and the man, and had to say a lot of sorry-about-thats. Fortunately, the man was very hunderstanding, but he shouldn't have had to be. After that, my lead was clipped on and I was Marched Off The Beach in disk-race. Mum had a very red face from puffing up the beach, and then having to do haplogising. I are quite very actual in the Dog House. Big-Time-Badly.

Now, it is a quite very well-knowed fact that I are a friendly boykin to most doggies, but that is beside the very point haccording to Mum. Some dogs do not want to play and some dogs aren't allowed to play cos they is getting over a hoperation, and need to stay on a lead and just do pottering and getting some fresh airs blowed up their noses and their bums. And some peoples don't want to have a soggy, sandy, panting Worzel Wooface oafing up to them at thirty miles an hour, hespecially when they has got earphones on and don't know wot is coming up behind them. Happarently, I must do betterer. Much betterer.

February 23

Mum has decided that my bad beach manners is All Her Fault and it's she who Must. Do. Betterer. She reckons I did the ignormous recall when I was on the beach with Scout and Dodger cos she was showing off to her friends. Well, not so much showing off but at least being on her bestest behaviour. Cos it's one fing to make a hidiot of yourself in front of complete strangers but quite an-actual-other to do it in front of people who you might want to talk to again.

So, Mum was concentrating on where I was and wot I was doing and generally keeping a close eye on who was further up the beach. When she did call me back, it was before I did spot hinteresting peoples coming up the beach. And finally, cos she didn't want to look a complete plonker in front of Dodger and Scout's Mum, her yell was loud and had a side-order of panic and please-don't-let-me-down-Worzel hidded in it. Scout's Mum probababbly didn't notice. But I did. It was an altogether Far. More. Hinteresting. yell than Mum's usual hefforts. And I heard it before I did bogging off on my next hexciting hencounter. So I did decide to come back in case Mum's leg had falled off or she had founded a secret stash of sausages.

February 24

In most cases, Mum deciding that somefing is All Her Fault is a good fing. Cos then it does usually mean it is not my fault and it turns out that I are not in disk-race; she is. But just because somefing is not my fault, does not mean it isn't going to be my problem.

All day today we has binned practising 'come.' Well, p'raps not All Day but on at least three blinking hoccasions when I has binned having a snooze, Mum has called 'come!' And after the first time, when there was a bitta cheese that needed eating, I actual fort I had better come a bit quickerer. Bitsa cheese do not like being lefted on their own for very actual long. And in this house, with Frank lolloping about, they generally doesn't stay lonely for long: he eats them, and hinside a fat cat is not a good place to store my cheese.

After the third time, Mum did fankfully come to the conclusion that I are exerlent at 'come' in the house. And that I know wot 'come' actual means. Either that or she did start to fink I would get a bit bored with the hole wake-up-come-down-stairs-appear-hinterested-glare-at-Frank-eat-cheese game.

Frank is not bored with the game. A-very-tall. Frank has got this hole come-for-a-bitta cheese fing sorted in his brain, and by the third time he was racing down the stairs to get to the cheese before me. And doing sit. Hand adding a high-pitched wail to the performance in case Mum hadn't noticed him.

February 25

I are quite actual pleased to say that Mum has decided that I can do 'come' hindoors, and after yesterday's performance with Frank, she has decided that for the sake of his belly we should find somewhere else to do our practising. Today, we tried it in the garden. Frank was most hoffended and sat on the windowsill in the kitchen waving his paws about and scrabbling on the glass, trying to get out to join us. Now Mum finks she's got a cat with a talent for hobedience. She hasn't: she's got a cat who is hobsessed with small pieces of cheese and with nuffink better to do than chase around after her.

The first two times Mum said come, I did actual hoblige but the third time I had other much more himportant fings to do so I hignored her. Kite, my fabumazing Labrador friend from across the road, came out of her house for a trip in her car, and I was quite actual keen to let Kite know I had seen her, and then to do watchering her get into the car, wot was not helegant, and listen to Kite's Mum say a lot and a lot of words beginning-with-B about honly having a two-door car, and next-time-I'm-having-four-doors-heven-if-it-isn't-cool.

So I was not very hinterested in come or bitsa cheese, and did demonstrate the hole problem: I will do come so long as I haven't got anyfing betterer to do. And no amount of Mum being hexcited and interesting or bitsa cheese will ever, hever be as hinteresting as Kite.

February 26

Just when I fort fings couldn't get any blinking worse, I has discovered that I are a poster boykin. It's a disaster and it's all Auntie Charlotte's fault and I are Not. Speaking. To. Her. Somehow, I has ended up with my picture on a poster wot is all about Good Dog Manners on the beach. Like I are a fine hexample of wot all doggies should do. Currently I aren't, hespecially the bit about not barging up to dogs on leads wot might not want to say hello. Mum says I are going to have to be heven more carefully managed from now on. Very. Carefully. Managed.

february 27

I aren't quite very actual sure I does want to be managed, and tonight Dad did begin to actual agree with me, and did wondering why me running up to other dogs is such a fuge problem: all I did want to do was say hello, and I was a perfickly friendly boykin. Which is all true. I was hopeful for a moment that Mum might See Sents and do actual hagreeing with Dad, but it seems not. There was a lot and a lot of talking about some dogs being scared by a blinking great oaf of a Lurcher, and others gettering over hinjuries and hoperations, and heven more stuff about it being Ill-Eagle to have a dog dangerously-out-of-control. Mostly, though, it was a lot and a lot of stuff about Worzel being on that blinking poster and getting a Bad Reputation. And Mum getting a worserer one. And if we don't get it sorted, Worzel Wooface will never, HEVER be going to the beach again.

So now I are really, wheely Not Speaking to Auntie Charlotte OR Mum.

february 28

Dad has dunned saving the day! And Worzel Wooface from never again being actual allowed off his lead to do zoomies on the beach. It's simple, really, he says. Beach-walking and marsh-zooming will have to be dunned when he can come, too. That way, there will be two of them to watch me and make sure that I aren't about to do bogging off or hinterfering with other doggies. Hand Dad has broughted me a noo toy wot is honly going to come out when we go to the beach. In-very-general, I aren't much hinterested in toys: I don't have a favourite but I do like noo ones a lot and a lot. So Dad reckons, if I honly hever see this toy at the beach, I will fink it is a noo one and want to play with it.

Today, we did try it out, and although I are still going too actual far for Mum's likering, Dad and Mum flinging the noo toy to each actual other was quite actual hentertaining, so I did not hinterfere with any dogs on leads or knock over any peoples listening to music in their Hearholes who were not lookering where they was going.

It's not perfick, Mum reckons, and I will always be a bit of a Noaf and run really, wheely fast, but it is bearababble. So long as we go to the beach at quieter times then fings will probababbly be quite very actual okay.

MARCH

March 1

The previously ginger one says she wants to save the world again and do some actual more with her Men-Tall Elf campaigning. Mum says she finks the previously ginger one should concentrate on saving herself for once but, happarently, if the previously ginger one saves the world, she will also save herself as well. She says she wants to Do. Somefing with her Naward for Houstanding Bravery she won at the end of last year, rather than just let it sit on a shelf being booful. Mum says that is habsolutely fine as long as Mum can have a lie down before any saving the world starts: she is quite actual tired, and she's got a Ned Ache.

March 3

Now that we has sorted out the Rules for Worzel's beach runnering, and the hole being off-lead stuff, Mum is feeling most actual relieved. How-very-ever, Dad can't do coming for every walk I need or be there to hold Mum's hand all the time. Fankfully, Kite's Mum has dunned coming to the rescue and has said that I can come round to her house with Mum when I do need to have a run.

And that means that Kite will get to have a run as actual well when Kite's Mum is too actual knackered after work. Teck-nick-ally, Maisie, Kite's older sister, will also get a run, but as Maisie honly seems to be able to run for ten seconds before clapsing on the floor or wondering where her dinner is, she could do that run from one end of my hallway to the other and the hole of Kite's garden is actual wasted on her.

Mum and Kite's Mum has also decided to leave all the runnering to me and Kite. For some reason, they seem to fink that cos they have no talents at runnering, there is no point in heven trying. At least Maisie has a go, and I do really, wheely fink they should stop all this lazy-Moo-Maisie and go-outside-and-play-Maisie talk.

March 4

Kite's Mum has got a Summerhouse. I has No. Hidea. why it is called a Summerhouse cos in the summer, everyboddedy sits outside. They does honly really huddle in there when it is not summer, so I do fink it should be called a Winterhouse. How-very-ever, I has decided I will not huddle in the Summerhouse, heven when it is winter. Quite very sometimes, though, I will wiggle in for a quick hello-luffly-boykin, but it isn't a place to do hangering around in as it is Not. Safe. for Worzel.

It's Not. Safe. for anyboddedy, really. The chairs are all comfy and warm, and they does look safe a-very-nuff, but when Kite goes in there with a tennis ball, noboddedy really wants to get up and go outside and frow it for her. Kite can be actual quite hinsistent about the hole tennis ball fing, though, and heventually somboddedy will take pity on her and throw it out of the door.

Or not. Mainly not, to be quite actual honest. And then the tennis ball pings around the quite very small Summerhouse and bounces off the wall, or comes straight actual back at the person who did frow it and smacks them on their nose! And then there is a lot and a lot of yowling and howling, and it is quite very actual Not. Somewhere. I want to be.

Some of my favouritist peoples in the hole wide world sit in that Summerhouse but that is cos they is stoopid. Probababbly from being hitted on the head with a tennis ball too many actual times. It is quite very possibibble for someboddedy to be a favouritist person and stoopid, but it is not possibibble when they is in that Summerhouse when Kite brings them a tennis ball. Then there is actual nuffink they can do or say wot will get me in there. Because I aren't stoopid.

March 5

Dad's rope naddiction is back. He loves ropes so actual much that he has decided to work out how to do the washing machine. He says he's always known wot it does; just not how to make it do it. But for a bitta rope, he was prepared to be brave and Press. The. Buttons.

He did ever so quite actual well. He pressed all the buttons and the washing machine did going round and round. And the special bitta rope survived as well. Now all Dad's got to do is work out which packet is soap powder and which one is stain remover, and realise that he probababbly doesn't need nearly as much conditioner as he did put in. In very actual fact, ropes probababbly don't need conditioner a-very-tall. Fings got a bit, erm, bubbly, and now the ropes are actual quite smooth and slidy, which none of us is sure is a good fing on a boat. But it's a start, Mum says.

Hexcept now she has got to clamber over a fuge piles of ropes to get to the pile of washing that needs doing, and it's all far, far too much like blinking hard work. But as she has dunned showing Dad hexactly where all the bubble-making stuff goes, he can have a go at putting clothes in the machine as well as ropes.

March 6

It has binned raining here all blinking actual day wot I are not keen on. A-very-tall. So, when it did finally stop raining for a little moment, I did have some fuge zoomies with Kite and Lola who is staying with us. After I did zoomies, I did realise it was my quite very actual dooty to do finking ahead, and so I eated lots and lots of grass. And I would not stop doing it, heven though I was asked several times, and tolded I was grazing like a moo-cow.

So now I has gotted a belly full of grass, I are ready to do providing the hindoor hentertainment cos the rain has started again. I does not know how long this rain will last and I fink it is my himportant work to make sure Mum does not get bored. Cos she can't do gardening.

So, now she can spend the hentire hevening following me actual about so I don't yack it all up on the carpet. I are quite very kind and fortful like that I do fink ...

For THE QUITE _{very} actual LOVE of Worzel Wooface

(handwritten: "very" inserted above "QUITE")

March 8

I has had some words made up in my onner. Happarently, I are now a verb, and that is a doing word like to run or jump or chase. There is now a hole noo verb, which is To Worzel. The peoples in Kite's house say fings to Mum like would-you-like-to-Worzel? and are-we-Worzelling-tonight? Like it is some super hactivity that they does do. And it is a super hactivity for me. But not for them as no hactivity goes with the words a-very-tall –

****TO WORZEL****

- There is halmost No. Doing. At. All. by any of the hoomans wot claim they are 'Worzelling'
- There is a lot and a lot of sitting round
- And drinking cuppateas
  The most 'doing' anyboddedy who is Worzelling does do is try to get the Summerhouse key to turn
- 'I can't Worzel tonight' is not much fun for me, but if Dad finks Mum is missing out on 20 minutes of brisk hexercise he is very missing the point
- The honly time anyboddedy gets any hexercise is when Kite's Mum rants about a useless-dimwit she has hencountered at her workplace. Then her heart starts to beat too actual fast and she has to sit down and have another cuppatea until her breathing does go back to normal again
- Worzelling sometimes hinvolves Wine on Friday nights
- And sometimes other nights if there has binned more than one useless-dimwit at the workplace.
- Teck-nick-ally. Dad and Kite's Dad are hinvited 'to Worzel' but the only additional hactivity that this will lead to is a bit of budging up on the sofa
- All this Worzelling is making Mum's bum grow, which might count as 'doing' but is not somefing Mum wants it to do doing

March 9

The previously ginger one says that she is going to save the world by doing a walk. Gandhi says he would be very glad if in future the previously ginger one did not say fings like this when Dad has got a mouthful of tea, and hespecially when he is using Dad's lap as a sleeping cushion. Cos now he is quite very actual not asleep but very wide awake, and also sticky with spatted-out cuppateas.

Dad reckons the previously ginger one might need to refink her hole walking plan: he's never seen her walk further than the kitchen without needing to get in the car ...

March 10

This weekend I are going to be Dad's dog. I does mainly like being Dad's dog cos it do hinvolve quite a very lot of snoozing on the big bed. And chips. Before we get to the weekend and the bed and the chips, I are going to do Management in the Hoffice at the boat yard. Dad said last night that I are such a growed-up boykin nowadays, he finks I won't be a bovver any-actual-more at work. Mum reckons Dad has binned Lulled. I hasn't got a clue wot Lulled is but I fink it's got somefing to do with False Scents of Seker-a-tree.

Dad isn't allowed to smell the Secreetaries at his work, Mum says. I are allowed to do this fing and, as far as I can remember, they does smell of soap, but I will do checkering this morning. And also try to remember not to do whining too actual much when Dad does disappearing, or Complaints to the Management if Lola isn't actual there to keep me comp-knee. Mum is going to Crufts this weekend to do working for *Dogs Monthly Magazine*, and I aren't allowed to go cos I aren't a peddy-gree. Or super-hobedient. And I are quite very actual Not. Doing. Dancing.

I does fink, all-in-very-all, I are quite pleased I has not binned hinvited.

March 12

Today I has got Complaints to the Management and I does hope Mum comes home quite actual soon. Dad and the previously ginger one are not actual any good at clean-as-you-go, and the house is starting to look like a Salt Course with added Cups; we're running out of chicken wings and, currently, I are the honly one who seems to have noticed that Mouse has dunned a poo behind the sofa ...

March 14

Mum is home and she did bring raw chicken wings, and she has dunned finding all the cups and dealing with the Salt Course ... but not the poo behind the sofa. She made Dad do that cos otherwise he will Never Learn, happarently. Or he will learn to be out when Mum comes back from a trip away so she sorts out everyfing and mutters about him being useless. Behind his back. Rather than right to his front.

March 15

The previously ginger one's walk is going to be called the Warrior Walk and it isn't going to be a very long one, honly a mile. Dad wanted to say that it didn't sound very warrior-ish or warrior-ful but he isn't sure that either of them are real words. They aren't. The word Dad was lookering for is brave. But as he didn't know all the details – and also cos Mum gived him A Look and then a kick under the table when he started to try and talk again – he did just shut up and do exerlent listening to the previously ginger one's plans.

She says she is going to walk from one end of Lowestoft beach to the actual other With. Her. Arms. Out, and she is going to ask anyboddedy else with a Men-Tall Elf to do the same very actual fing. if they does want to; if they is feeling brave.

And then, Dad did hunderstand himmediately why it is going to be called the Warrior Walk and why it is brave, but I do fink I had better do some hexplaining because hunless you does live with a Men-Tall Elf, you may not hunderstand a-very-tall.

Men-Tall Elfs can make the people they live with feel really, wheely bad about themselves. Or sort of half-dead and with no feelings. When that actual happens, sometimes peoples do scratch or burn or bite or cut their arms to make themselves feel betterer. I has got No Hidea why it works and the trouble is ... it does work. And noboddedy really hunderstands why, which makes it heven more difficult to stop doing it and find a safer way to feel betterer.

So, the previously ginger one has got a lot and a lot of scars on her arms that she is actual quite shamed about, just like a lot of peoples with a Men-Tall Elf. She should not really feel shamed, but it is all part of the Fighting the Stigma of Men-Tall Elf.

When it is hot in summer she wears long sleeves on the beach because she does not want peoples to see her arms and fink she is a bad person. And she says it would be nice, just for one day, to wear a short-sleeved top and feel the sun on her arms. And if everyboddedy who has a Men-Tall Elf and scars wot they keep hidden gets them all out on the same day, in the same place, and does the same walk, they will have some comp-knee, and won't feel like they is so terribibbly alone.

Dad is possibibbly, probababbly halmost certainly quite very glad Mum kicked him now.

March 16

There is a really, wheely tasty bitta grass on Kite's lawn, and Mum says I are hobsessed with it. Noboddedy can remember anyfing being dropped there, and as I has binned henjoying it for weeks and weeks now, anyfing that was dropped there that might have binned tasty would be long gone.

At first, noboddedy did notice my hobsession but, like lots of actual fings, once you start to notice somefing, you can't *un*-notice it. My grass sniffing and nibbling is now being very actual hobserved and commented on every blinking day. Today, Mum kindly watered my special bitta grass wot I do fink was quite very kind of her. She finks she is washing away anyfing wot might be tucked up in the grass so it doesn't do hurting me, but you will be quite actual pleased to know it is still just as fabumazing as it was before. Just wetter.

I DON'T LIKE MONDAYS

Kite's Mum does singing on Mondays
She has to go in her car
There's just about time for her dinner
Before she does tra-la-la-la

There's no time for playtime with Worzel
She's got to straighten her hair
But I doesn't hunderstand Mondays
I don't fink that they're very fair

Every day Dad comes home from his work
And then there's a cuppatea
And a chat with Mum about how was your day?
Then it's playtime for Kite and me

But that doesn't happen on Mondays
Mum doesn't stand up and say
You coming? To me as she picks up my lead
Like she does every other day

So I really don't like blinking Mondays
Toy playing won't do instead
I want to see Kite like I always do do
Mondays confuddle my head

March 17
****FUGE NOOS****

Pip and Merlin have dunned moving back to my village. And they has binned
to visit me! Has you ever metted someboddedy who you just LOVE ... really,
wheely love, even though you does not live with them? I does feel that actual
way about Louise, Pip and Merlin's Mum. I do trying to sit on her lap and give
her kisses. It is very quite confuddling cos there are other actual peoples I do
see all the time who I do not try to do this fing with. Mum reckons it is cos I has
knowed Louise since I did first come to live here, and all my memberries of her
are luffly and stucked in my brain. Also, she did bring Merlin with her, who is my
oldest friend wot will always get my actual happroval, and Pip, who is always
hentertaining!

Mouse does not feel the same way about Pip and Merlin coming to
visit. In very general, Mouse does not mind Merlin too actual much because
he is a cautious boykin, and he sticks very quite close to his Mum in case she
does decide to leave without him. Pip, on the other hand, finks our house
is fabumazing cos it is designed for ignormous Worzels wot can't stand on
windowsills. The best I can do is to stand on the sofa and put my two front
feet on the windowsill, but then I are kinda stucked in that position, and
getting down hinvolves a confident backwards jump that mustn't be fort about
too actual much. If I fink about it for too long I does find myself heven more
stucked, and the confident backward jump becomes a tenty-tiv hop followed by
a hunexpected smack on my bottom jaw from the windowsill. So I try to havoid
it.

But Pip can skip and skitter up onto the windowsill just like a very small
mountain goat, and today she managed to get off the windowsill and then
onto the shelf where Mum keeps her special fings. And cos Mouse finks she is
one of those special fings, she does often lie up there. Once Pip made it onto
the shelf, Mouse decided she had no choice but to find somewhere else to be
special, in a very quite fluffy, hysterical way. And she was so busy giving Pip
disgustering looks and making sure Mum and Louise noticed her sulky door-
slamming-dramatical-stomping-out-of-the-room, that she completely forgotted
about me and Merlin.

Which kinda spoilt the effect she was trying to hachieve I do fink, as it is
actual very hard to stalk off in a disgusted-but-digger-ni-fried way when you has
suddenly got two Lurchers chasing you out of the room ... and out of the catflap
... and into the rain, and a Pip frowing everyfing else off the shelf whilst she
does hexcited commentatering on the action.

This afternoon, Mum says, we will be rearranging the sitting room and
trying to persuade Mouse to come out from under the car. But mainly we are
very happy our friends is back.

For THE QUITE ~very~ actual LOVE of Worzel Wooface

March 18

For some reason today Mum did lying down on the carpet and shoutering numbers at it. I has got No Hidea wot is going on but I are very quite sure the carpet has not dunned anyfing to deserve this yellering, apart from being a bit actual grubby.

Then she gotted up and did carry on as if nuffink was wrong.

I fink Mum has finally losted the plot and if the carpet is gettering yelled at for being a bit dirty and smelly, then I are probababbly going to be actual next. I are going back to bed and staying out of the way.

March 19

Mum shouted at the carpet again today. I has still got No Hidea wot is going on, and why the carpet has hoffended her so actual much. Today's shouting session was a bit longerer and I fink the carpet has started to fight back cos Mum did not stand up and walk away as if nuffink had happened. She staggered to her feet and had to do leaning against the door frame for a bit.

I are still keeping out of the way, but currently there do not seem to be any plans to do shoutering at me. Or give me a bath.

March 21

Mum says I has got to stop eatering that patch of grass. She says it can't be good for me, and wot is heven stranger is that I are not yackering it up like I do usually do with grass, so there is hobviously somefing else going on. And also I are not a sheep.

March 22

I has discovered wot is going on with the hole carpet shoutering stuff. The carpet has dunned nuffink wrong: it's just in the wrong place at the wrong time. Mum has decided that she needs to do somefing about the size of her bum, and also do somefing to stop her back hurting. Mr Google on Uff the Confuser says she should do somefing called the Plank Challenge. It's all about getting a Cor! Happarently.

Currently, Mum does not have anyfing anywhere in her boddedy that would count as a Cor! She reckons it's more like an Ughh! and the hole idea of the Plank Challenge is to get less Ughh! and more Cor! To get a Cor! Mum has to lie on the floor resting on her elbows and her feets and count to a number. And each day the number she counts to will be higher. But she is not allowed to count faster. According to Uff the Confuser, in a month she should be able to count to 300, wot sounds like a fuge number. I don't fink Mum will be able to count that far without passing out or clapsing in a heap on the floor.

Mum says she is going to try, and the picture of the man on Uff the Confuser doing all the hencouraging is actual quite nice to look at whilst she is counting. He has smiley white teef and halmost no clothes and lots of big, shiny muscles. And that all adds up to a lot of Cor! And some Phwoar! as well.

Mum says she has No Hexpectations of getting to Phwoar! but a bit of Cor! would be nice.

March 24

Kite's mum is also carpet-shoutering-plank-doing. I don't fink they is having a compertishon with each other but they is being hencouraging, and every evening there is talk about have-you-dunned-your-planking-today? And then a lot and a lot of whining about how hard it is. I do not hunderstand wot is so difficult about it to be quite actual honest. I has had a good look, and planking does look like a stretched out play-bow to me. And I could stay in that position for hours if I did need to.

March 25

Mum and Kite's Mum are having mixed actual results with their planking. Both of them has managed to get to 50 counts, wot they is quite actual pleased about. Even though the carpet shoutering is nuffink to do with it being grubby, I are still keeping out of the way of Mum when she is doing planking. There is No. Way. I are getting muddled up with Mum lying on the floor trying to remember to breathe and shout numbers out of her mouth like they is the last words she will ever say. I are watching from the top of the stairs until she gets up. Once she has stooded up, it is my himportant work to give the carpet a good sniff to make sure that it has dunned surviving being shouted at, and then lie on top of the hexact spot Mum has just yelled at.

In Kite's house, they is doing fings differently. Kite has decided that her Mum is either in very deep troubles or in need of hassistance and hencouragement. As soon as Kite's Mum lies on the floor, Kite finks it is hessential to give her Mum licks, and she snuggles up to her doing worried little wriggly sitting down tail wags. Kite's Mum says it is himpossibibble to do planking in the same room as Kite so, from now on, Kite has got to go outside when she is planking: she can't plank and count and breathe when she is also having to spit a Labrador's tongue out of her mouth.

March 26

Dear Kite's Mum

I would like it to be actual knowed that it was Not. Me. wot stole the cold sausages out of your fridge, and it was Not. Me. that eated them. Well, I might have had the nubby little end bit that's crunchy and covered in yummy fat. And Kite and Maisie might have had an ickle bit as well. But we did just actual receive stolen goods, not do the stealing. Or even ask for them to be stoled on our behalf. Us dogs are Not. Guilty. of any quite actual crimmy-nal acts.

It was Mum wot spotted them when she went round to let out Kite and Maisie at lunchtime when she opened the fridge to get the milk to make her cuppatea. And it was Mum who was hungry and who has got no self-control about cold sausages, and also Mum who did not do knowing that they was a vital hingredient for your dinner tonight.

From your very actual shamed luffly boykin

Worzel Wooface

Pee-Ess: Mum says she has got some nice ham here, and would that be a fair swap?

March 27

The mystery of my grass-eating has binned solved and I are a very clever boykin! I has binned eatering wheat grass that some birds did plant for me a long very time ago when there used to be a tree growing in the lawn, and they did droppering the seeds into the soil. Kite's Dad sawed me eating the grass there tonight, and did actual remember the tree and putted all the clues together.

Happarently, wheatgrass is a superfood and it isn't going to do me any actual harm, so I are allowed a little but not too actual much. I do still look like a sheep, according to Mum; just a clever sheep. I don't fink there is any such actual fing as a clever sheep but if it does mean that I can nibble my special grass hoccasionally, I are happy to be the first and honly clever sheep in Ingerland.

March 28

There is Lions in Lowestoft I has discovered. Fundreds of them. And these Lions are going to do helpering the previously ginger one with her Warrior Walk. There has got to be somefing very actual wrong with all of this but currently noboddedy else has noticed. Mum keeps saying fings like oh-the-Lions-are-brilliant-at-stuff-like-this, and is very actual glad that someboddedy other than her is going to be In. Charge. cos she hasn't got a clue about collecting buckets and stopping traffic but the Lions do it all the time, she says.

To be quite actua honest, if I saw a Lion walking down the high street in Lowestoft and it was collecting buckets I would hand over my bucket straight a-very-way. And if the Lion fort I was traffic and not a dog, I would do whatever it blinking well wanted me to cos they is cats, honly blowed up twenty times biggerer.

I find it hard a-very-nuff dealing with the normal-sized cats wot live here, and when I meet a strange cat in the street I do try to remember to give it a wide birf, and then woof at it. Or woof at it and then remember they has claws and I has bitter hexperience and then do givering it a wide birf. It depends which bit of my brain gets itself horganised first, really: the woofing or the bitter hexperience.

But if I saw a Lion, I don't fink woofering or bitter hexperience would be a lot of help. And I don't actual care if it is a Warrior Walk, I are planning to run away, very quite actual fast and not at all bravely ...

March 29

I has made a noo friend this week. Hunfortunately, I has not founded out wot his actual real name is but I fink he is related to the Easter Bunny. He is quite actual friendly and wise, and heven though he honly has one eye and is very, very senior, he is quite actual good at choosing places to sit where he can watch me being a hidiot without getting barged into. And also watch me find all the presents he has lefted me on his bit of grass behind the boat shed.

Happarently, honly senior and very quite special cats can do this fing. It is their special skill. I aren't allowed to tell you wot sort of presents these is cos Mum says you will know hexactly wot I mean. And so does she, now. I has rolled in them, tried to eated them, flunged them about and ... it's been just like

Easter, honly for dogs. And the bestest fing of-very-all is that hoomans can look and look and LOOK for these presents and they never see them. No matter how hard they try. Well, not until they is stucked all over my collar. Then they can find them.

March 31

It's going to be Gipsy's birfday in a couple of days and we are halmost certain she is going to make it. She will be 13. That isn't very old for a cat but for Gipsy, it is a miracle. Ever since we did discover that she has somefing wrong with her breathing we have binned waiting for the day when the medicines don't work anymore, or they did start to do her more harm-than-good. It's all been about qualerty of life, and making sure she can still be the cat she wants to be.

Trying to persuade Gipsy to be a cat she *doesn't* want to be would be pointless and painful and very actual lonely for her because whenever anyboddedy tries to get Gipsy to do anyfing she does not agree with, she just slides out of the catflap and we don't see her for days.

But currently, she is coming in each day for her breakfast and her dinner, and we have put some chairs and boxes around the kitchen so she can climb onto the table without having to jump. Dad reckons it's like having a very quite too-hindependent-and-mad-old-lady living with us: you make little changes and lie about why you really, wheely want to do the cookering when all you really want them to do is to sit quietly and not burn down the house.

We are all hoping we is gettering away with the helping-not-helping: if anyfing changes too fast, or if anyboddedy even finks of helping out Gipsy in a way that she notices, she'll bog off big-time-badly in a sulk.

APRIL

April 1

The fuge ginger boyman phoned us tonight. His Confuser has died and it is a disaster and he is going to Fail! His! Degree! he says. Halmost everyfing the fuge ginger boyman says starts with a big letter and ends with a hexclamation mark. Heven down the phone.

Sometimes, when I fink of the fuge ginger boyman, I do wonder how Mum managed to produce anyfing so fuge and, well, ginger, cos it does not make any sense a-very-tall. Mum is not fuge or ginger so I do not know how this happened. Happarently, it has somefing to do with jeans again, and there must have binned a lot and a lot of peoples in our famberly who is now long dead who had ginger hair. But you can't tell cos all the photos are black and white. And some of the peoples wot are still very actual alive did were ginger before they discovered hair dye. Now I are wondering if the hole hair dyeing fing is also about jeans, and maybe that is why the previously ginger one is always doing it.

Anyway, at other times, like today, I does have no problems a-very-tall believing that the fuge ginger boyman did get borned by Mum, cos he did show off lots of his famberly jeans by being quite very loud and having actual hystericals about this blinking Confuser. And did making such a fuge panicky fuss that Mum couldn't stand the noise any more, and she has dunned actual agreeing to hire him a laptop Confuser, wot is smaller than a normal Confuser, and send it to him so he doesn't fail his degree, and so he can finish his diss-the-nation. But mainly so he'll stop using so many words down the phone.

Dad says the fuge ginger boyman also seems to have got Mum's famberly jean talent for getting-your-own-way. And then dunned using it against her ...

April 2

Dear Mum

No, I does not know wot is in the actual hedge, which is why I are watchering it. And I are never going to actual find out wot it is if you keep distractering me with your wot's-in-the-hedge-Worzel? words. Now, please will you shush up for a blinking minute so I can do finding out.
From your luffly boykin
Worzel Wooface

Dear Mum

No, I does still not know wot is in the hedge. Every time I do fink I has worked it out you start saying Wizzy-Woo-come-away and distractering me. I fink it is a live fing, though, and if I can just do stretching a bit actual further I will do finding out.
From your quite very actual determined boykin
Worzel Wooface

Dear Mum

I does not know what a proverb is, and saying a bird in the hand is worth two in the bush is actual quite losted on me. And anyway, you does not have a bird in your hand, you has got a lump of cheese. And now, after all your wafting of cheese and wailing of words, there is no blinking bird in the hedge. It has dunned bogging off. So I will do having that bitta cheese now.

From your not begging, not choosing, not looking a gift-bitta-cheese-in-the mouth luffly boykin
Worzel Wooface

April 3

Mum has hired a Confuser and she has sent it to the fuge ginger boyman. She has dunned very quite well. Halmost. All the making sure it had all the right bits of kit on it did go very actual hefficiently, and the queue in the post office wasn't too long so that bit went perfickly, too. The part where you write the address on the parcel didn't go so well, though ... Now the parcel is on its way and, haccording to the very patient man in the big sorting office at Halesworth, it's somewhere on a motorway or on a train heading to the right town in the right county, but to the wrong flat, in the wrong block, and no, he can't stop it. That's not how the post office works, and stopping the mail is ill-eagle and used to get people's heads chopped off, and he wouldn't recommend it.

Mum finks it might be a better hoption to be actual quite honest but now she just wants to cry. And drink wine.

The fuge ginger boyman did not do saying any words beginning with B to Mum about this. He didn't say nuffink. Stunned Silence, it's called and heventually, when the peoples in my famberly get over this disaster, they might realise that this was a sig-niffy-cant moment. In the meantime, Mum tooked hadvantage of the fuge ginger boyman's Stunned Silence and said more than a-very-nuff for them both. She did call herself so many actual rude words that by the time Dad got home, there was none lefted for him to use. So he just gave her his bestest scay-fing look which is worserer, apparently.

So then everyfing got quite tents and I did decide to stay hiding up on the bed until it was time for my dinner. And then I skulked past the hoffice hoping that the scay-fing would do going away. There was heven a small moment when I did fink I would not be able to eat my chicken wings.

But I are pleased to say that the previously ginger one did saving the day – hand my dinner. She wants to know when Mum is going to buy her a new Confuser because It's. Not. Fair. And that did cause a small-but-very-you-knighted parental hexplosion.

And now the horrid-and-very-quiet hatmosfear has gonned, I are pleased to say I can finish my dinner in peace. Well, my famberly's version of peace which is quite actual noisy, and not the Orrendous quiet that has binned floating around for the past three hours.

April 5

I just opened a door All By My Own. It was open just a-very-nuff for me to be able to do that fing. I did using my nose and my paw, and pulled it towards me until I could do fitting through the gap! I did it all on my own without having a panic or waiting for Mum or Dad to do helping me. And I didn't know wot was on the other actual side neither. It might not sound quite very brave, but for me, it was a fuge big fing.

****FORTS ON DOORS****

- In very general, I would prefer it if doors didn't hexist, as doors are much scarier fings than hoomans fink
- Door live in narrow spaces. When fings get narrow I have less choices about running away or turning round. Going through a doorway requires commitment. I aren't always prepared to do committed
- When doors are shut I can't see wot's on the other side. It is my hessential and himportant work to shout as loud as possibibble so that the Fing. On. The. Other. Side. is more scared of me than I are of it. As it is halmost himpossible for the fing-on-the-other-side to be more scared than I actual am, I have to be really, wheely loud. And hope it does running away
- When doors are open, they waft about and sometimes slam shut in the wind, and that is terry-frying and could do killering me
- Finding myself on the wrong side of a door is a disaster. My hidea of the wrong side might not be the same as yours; in actual very fact, it often isn't
- I are quite very actual sure there must be a worserer crime than leaving a door open at Gran-the-Dog-Hexpert's house but what it is, I aren't sure. And neither is Mum who has knowed her for-Hever
- Some peoples used to say that hoomans should always go through doorways before their doggies because it showed the doggy they was the 'pack leader.' This didn't show their doggy nuffink of the actual sort, and if you does want to be a pack leader, go and join the Scouts, they is quite actual desperate for them
- How-very-ever, it is still a Good Hidea and very quite kind to your doggy to go through the doorway first so you can protect your doggy from woteyer is on the other side. Even if it is honly a crisp packet blowing in the wind
- The vegetababble most commonly eated by dogs is a door. And it isn't for the vitty-mins

April 6

Gipsy has got a hole in the back of her neck. We aren't sure wot it is but it could be her skin getting very actual thin from all the hinjections and medicines that she has binned having. Mum has put some cream on it to see if it will heal up, and fortunately, it isn't in a place where Gipsy can lick it off. She couldn't heven get it off when she struggled out of the towel Mum had wrapped her in. And it didn't come off on Mum's jumper when Gipsy scratched her big-time-badly on the side of her face, neither.

Now Mum and Gipsy are both stomping around the house with white blobs of cream on them, looking equally not-very-pleased about it ...

April 9

Today, Mum is wondering how she is going to tell Auntie Di in the hiring shop that she has losted the Confuser that she did hire to Mum. Dad says Auntie Di will probababbly want back her telly and her dishwasher before Mum sends-them-on-a-flipping-magical-mystery-tour-to-Scotland, and there is Fat Chance and also No Way he is calling into the hire shop to confess Mum's sins on his way back from work. Then he stomped out of the house and had a fight with the wheely bin and slammed the car door. I fink he's still quite cross.

I are not cross with Mum. I did wot any other dog would do in this situation and hinsisted we did go for a distractering walk, and hencouraged her to stomp about and do some finking. By the time we got back from our walk, Mum had very muddy legs and a plan. It did just rely on her getting her wellies off and then finding out the phone number of the sorting office in whatever bit of Berkshire would get the parcel.

The post office isn't actual designed for peoples to phone them, which actual makes sense when you fink about it. If everyboddedy used the phone then there would be no point in having letters, so it isn't somefing the post peoples would hencourage or make easy ... but heventually Mum's hinsisting and button-pressing got her a phone number and, for a tiny actual moment, Mum fort all her problems would be solved. The man on the phone in the sorting office in Barkshire saw Mum's parcel with his actual hiballs. Everyfing should have dunned ending happily, but that ill-eagle is still causing problems. And Mum still can't have the mail stopped, no matter how much she wails. And Mum could be anyboddedy. Heven the man on the end of the phone didn't believe that, though, and did very actual hagree that honly someboddedy who had put the wrong address on a parcel and got Stunned Silenced at by her hentire famberly, and then spended an hour pressing zero-to-speak-to-a-hooman, and then describing the blue plastic stuff around the outside of the Confuser in fuge detail could possibibbly be the person who posted the parcel in the first place.

And that was all quite very actual well and good but it didn't matter. Nuffink is as himportant as ill-eagles, it turns out, so the postman has got to try and deliver the parcel or he really, wheely will get into fuge troubles, heven if they don't chop off people's heads any more. So now all the fuge ginger boyman has got to do is to find the flat and then get them to hand over the actual quite hexpensive parcel, or the ickle card wot says the postman couldn't deliver it. And then go and collect it from the main delivery hoffice which is 20 miles from where the fuge ginger boyman lives, and he hasn't got a car. Or any time. And also he's got a diss-the-nation to finish and a deadline and still-no-blinking-laptop.

As you can probababbly tell, there was no Stunned Silence this time.

April 10

Gipsy's hole in the back of her neck is starting to heal up. Mum says this is a fuge relief, but she doesn't know if Gipsy will be able to have any more steroid hinjections any more, or if it was caused by the fleas drops or whether it was the two fings reacting against each other. Either way, Gipsy is going to need to see Sally-the-Vet.

Mum fort it was about time all the other cats got a check up as very well, so instead of Mum taking all the cats to the vet, Sally-the-Vet is going to come and visit us. Mum says that by the time she has found, catched and shoved five cats into a cat basket, then driven backwards and forward to the vet five times, it will probababbly be much cheaperer and a lot easier just to lock the cat flap and keep them in for a morning. Dad reckons Mum should borrow some cat baskets and transport them all in one go to the vet, but I do not fink he has properly fort out this fing, and perhaps he has forgotted that we does have a small car. And heven if we had a fuge stretchy limmy-zeen, it would not be actual big a-very-nuff to transport all of our cats to the vet. They all actual hate going in a cat basket, and yowl and howl and try to escape. I don't fink Gipsy has ever actual stayed in the cat basket for a hole journey, ever. And it's only two miles to the vet.

Mum says she doesn't fancy driving with five cats all trying to make a bid for freedom all over her windscreen, and what if she gets stopped by a policeman? Mum says Dad is quite very welcome to try, though, and she'll come and get him letted out of the police station once she has founded all the cats that would scatter like the wind as soon as he opened the car door ...

Dad has now decided that Mum's plan is probababbly a good one.

April 11

It does look like everyfing might be actual okay with the laptop Confuser disaster! Maybe ... The fuge ginger boyman has looked on a map for the address wot Mum put on the package, and he does know the man wot lives there. Dad says that as the fuge ginger boyman does live in a fuge town, this is a proper mirror call, but then he doesn't know how the fuge ginger boyman knows him and it is all Orrendous. And that just because you know someone, doesn't mean that you want to know them again.

A long, long time ago when he had just moved in to his noo flat, the fuge ginger man did too much drinkering Cider. He stumbled home to his noo flat and, for some hunknown reason, his key would not work in the door. After a bit he felt quite very actual tired so he did give up and decide to try again when his brain was working, and sat down in the doorway to have a little sleep. Like you do when you are 21 and it's two in the morning and you is full of Cider. The next morning, the fuge ginger boyman was woked by his head hitting the carpet with a fud when the flat door opened, and a Complete. Stranger. was staring down at him, wondering wot the Heckington Stanley was lying on his doorstep. And whether it was still alive.

Fortunately, the Complete. Stranger did remember the time when he was at Universally, and once he had dunned getting over the shock of finking that there was a dead boddedy on his doorstep, and realised that the fuge ginger boyman was just a bit cold, a lot actual hungover, and very, very quite sorry, they did both have a bit of an hembarrassed laugh about it. Then the fuge ginger boyman scurried away, and it was very actual hagreed-but-not-said that they would Never. See. Each. Other. Again.

And now, the fuge ginger boyman has got to go round to the same man's flat to get the card to take to the post office to get the laptop. Mum has tried to convince the fuge ginger boyman that it is all a Good Fing that it is this man, cos then at least there will still honly be one person in his town wot finks he is very actual Hinsane, rather than two. Or maybe the man won't remember him. And also can he fink about drinking less Cider in future ...

46

Heven Dad's gettering the hole hidea of Stunned Silence now.

April 13

The fuge ginger boyman has tried to visit the man in the flat he never wants to
see again but he hasn't been in. Either that or he isn't answering the door as he
knows who is on the other side. Dad has gived the fuge ginger boyman some
monies so he can go and buy another Confuser because he still has a diss-the-
nation to do, he still has a deadline, he still doesn't have a Confuser, and it is
wot Mum shoulda dunned In. The. FIRST. PLACE.

Me and the previously ginger one have dunned deciding to keep away
from all the noise. Neither of us is quite actual keen on yellering, which doesn't
happen that often here, fank very goodness. It makes both of us a bit actual
wobbly and hanxious, so we did hiding-in-the-kitchen, and the previously
ginger one cooked bacon. A hole packet. And she did sharing one slice of it
with me to take my mind off fings, and I did exerlent listening to her mutter
a lot and a lot more about how the fuge ginger boyman will end up with two
Confusers and It's. Not. Fair.

April 14

Today I has binned very quite firsty. Happarently, bacon can do that to you.
Fortunately, the cause of my firsty-ness was well knowed, and Mum says the
previously ginger one is old a-very-nuff to know that six pieces of bacon is not
good for her. But could she be a betterer hinfluence on Worzel and also tell her
next time. Cos she's honly just gotted over the last Die-A-Bee-Tees dramaticals.

April 15

Sally-the-Vet came to visit today. I was a luffly boykin, but when I sawed all the
weighing and measuring fings coming out of her truck, I did decide that apart
from a quick wiggle and a small lick, I would do hiding upstairs on the bed,
which is about as very far from the kitchen as you can get in our house and
today, being in the kitchen is not somewhere I wanted to be.

It is a quite very actual strange fact that when the cats are allowed to
come in and out as they do please, they seem to spread themselves around the
house and find quiet places to be. Gandhi lies on the previously ginger one's
bed; Mouse likes to find a warm bit of tecknology to lie on, and Mabel takes to
her shed. Frank, in very general, lives on the kitchen table during the day, and
in the room with the chair and all the water at night. Gipsy cannot be founded
in the house a-very- tall cos she honly comes in for her breakfast and her dinner,
and spends most of her time in the hedge up the road.

But as soon as the cats cannot get out, they all hang around in
the kitchen and do squabbling and arguing, trying to work out who is
responsibibble for the change in the facilerties and biffing the cat flap. I does
not know why they do this fing: under normal actual circumstances, none of
them would want to be outside at this time of day, apart from Gipsy, and she
was not joining in with Complaints to the Management in the kitchen, mainly
because she was locked away upstairs. Keeping Gipsy hinside when you
does not want her to go out is a hole nother hexperience: a bit like living in a

submarine and there does always have to be at least two closed doors between her and the outside world. It was also decided that Gipsy would not have a big check-up because she has binned checked up so many times recently that it would be actual pointless. And everyboddedy did agree that watchering all the other cats being checked might be too actual much for Gipsy to deal with.

So it was decided that Gipsy would wait until all the other cats had been dunned, and then, when they all shotted out of the cat flap in a fury, she would get a quiet stroke and listen to her heart and hinspecting of her hole in the neck in private.

I do hexpect that everyfing might have binned all very okay if Mum had not decided that she would also get Sally-the-Vet to do worming all the cats, and show Mum once-and-for-very-all how to do it properly. Without gettering scratched and bitted to shreds. At least that was the plan.

Mabel wented first with her check-up and, apart from finking that the scales were hobviously Not. For. Sitting. On. she did quite actual well. When it came to the worming, Mum said Mabel was the least of her actual worries, and Sally-the-Vet managed to get the tablet down her without too much hystericals before lettering her escape to her shed to get over the hin-digger-nitty of it all. Gandhi, I are actual quite pleased to say, tooked the hole worming fing in his stride, and then did wot any Onerable Lurcher would do – used his strides to do running away as fast as his legs would carry him, with his tail all puffed up acting like a fuge fluffy sail as he hurled himself over the fence. That lefted Mouse and Frank.

Wot isn't a great situation. Most people fink Mouse is a pretty cat. I does not know if this is true about peoples but, for a pretty cat, Mouse can make the most hugly faces of any cat I has ever seen. Perhaps it is because we is always used to her looking booful, but I has never, hever seen such an hugly cat as I did see in that moment. And she did do her bestest best to make Mum and Sally-the-Vet look hugly as well. With her claws. I do fink it was at this moment that Sally-the-Vet did decide that perhaps Mum wasn't being quite so actual useless about getting-the-cats-wormed after all.

So she did get out an Hinstrument of Doom. Well, maybe not Doom, but a not-messing-about-now tool: the Hinstrument of Doom did a clever trick of missing out Mouse's mouth and tongue and teeth and basically pinged the tablet right down past all that lot and into Mouse's stomach. I has got to say that Mouse's hugly face did change at that moment to very quite actual confuddled; as well as by-passing Mouse's mouth, the tablet going into her tummy by-passed her brain as well, so she wasn't actual sure why she was struggling and scratching, or why she was suddenly put down on the floor and allowed to escape. Mouse isn't wot you'd call a clever cat ...

Frank isn't particularly talented in the brains department, neither, but his digestive system is either hincredibly clever or a pain in the neck. Depending on where you was standing when the Hinstrument of Doom shot the tablet into his stomach ... and Frank's stomach shot it straight back out again and hit Mum who was trying to hold onto him.

Sally did wonder if perhaps she had made a mistake and had not dunned it right so she tried again. And Frank himmediately hejected the tablet again, wot is somefing that is not normally seen in cats. Though it is in Llamas, happarently.

At this point, it was actual decided that Frank would have to have a worming drop on the back of his neck instead of a tablet. And, fankfully, Mum got to put Frank on the floor before her arms falled off or he did his spitting Llama trick again with somefing heven more objectionable than a soggy tablet.

April 16

Frank is in fuge trouble. In actual very fact, Frank *is* fuge. He weighs nearly 9kg which is the same as a Beagle or a Shetland sheep dog, but not as much as a Llama. They weigh about 100kg, Mum says, but holding him whilst he was doing all the spitting and struggling felt just like trying to control a Llama, and if Dad doesn't stop feeding him so much, he can deal with the hole worming fing in future: it is redickerless and hembarrassing having a Fat Cat. Dad says Frank isn't fat, he's just big-boned and Mum-is-a-meanie-and-she's-hardly-one-to-talk.

So Mum has stopped talking. And cooking. But it's very okay cos Dad is making chips ...

April 17

The fuge ginger boyman has got the card for the post office! Hunfortunately, the man in the flat did remember the fuge ginger boyman cos you don't say "Oh-it's-you-again" to peoples you don't remember, and fanks to heven more of Mum's jeans, the fuge ginger boyman is not hexactly forgettable: he's fuge and he's ginger but, more himportantly, he can't get to the main sorting office, he says. They've told him they'll return it to the sender and I-assume-you-managed-to-put-the-correct-address-for-your-own-somefing-beginning-with-B-house. He has also broughted himself a noo Confuser with the monies Dad sent to him. And he's quite actual busy now. Until June.

April 18

Mum nearly kissed our postman. Fortunately, I did hear her squealing and being relieved and hexcited so I did saving the day. And the postman. All Mum's happy noises did make me happy, too, so I did give the postman a couple of tiny ickle licks on the hand, wot he was very quite pleased to get. I does have a strong feeling that postmen do prefer kisses from a luffly boykin doggy, wot are much more hacceptababble and normal than kisses from a Mum who is still wearing her dressing gown.

But the most very actual himportant fing is that the Confuser is back in safe hands! Well, safe if you does consider Mum to have a safe pair of hands, which halmost noboddedy in my famberly does currently. And also, honly actual quite briefly in Mum's hands if I are completely honest, cos as soon as the previously ginger one saw the Confuser parcel, she hannounced that she needs it, and as it's on hire for a year now, someboddedy might as well make use of it. And that It's. Honly. Fair.

Mum says she can have it: she never, hever wants to see the blinking fing again.

For THE **QUITE** ✓*very* ~~actual~~ LOVE of **worzel wooface**

THE TOOLIP
For days it stooded there
Tall and proud
A toolip in bud
Alone in a crowd ...

... of bluebells and daffy-dills
Pansies and grass
And the toolip stuck out
Flowering at last

And then I trodded on it
Lurching around
And squashed it quite flat
Into the ground

I'm a noaf, Mum says
And I'm very quite clumsy
It's a blinking good job
That she actual quite luffs me ...

April 20

The hole in Gipsy's neck has dunned healing up, but she did not actual happreciate Mum hinsisting on lookering at it. Last night when Gipsy went outside, she sat on the pavement for ages finking evil forts, and I was quite very glad there was a fuge fence between us. Noboddedy is actual keen on Gipsy finking: she is dangerous a-very-nuff when she does stuff without using her brains.

Mum says she wasn't finking, she was gettering her breaths back from climbing over the fence too fast ...

April 21

Gipsy didn't come in this morning, and Mum is not being very frilly-sofical about it a-very-tall. The last time she saw her was last night when she did hinspecting the hole on the back of her neck. And then once Gipsy did hescaping from Mum's clutches, she leaped over the fence and did the hole I-can't-breathe-fing before she stalked off in a huff. Well, not hexactly a huff because Huffs need Puffs and the one fing Gipsy didn't have was any Puffs lefted. Now Mum is convinced it is All. Her. Fault.

When Dad came home from work tonight, he was sended straight out to go and Find Gipsy. He did very quite badly to be quite actual honest, and when he came in and said I've-looked-and-looked-can-I-have-my-dinner? he did get told a lot of words about Mum not asking him to go and *Look* for Gipsy, he had to go and *Find* Gipsy, which was not the same fing a-very-tall. By this time, Mum was starting to have a quite actual wobbly face, and doing that hole fing when she starts to take fuge, gaspy breaths and flap her arms about. Neither me or Dad are any quite actual good at hunderstanding Mum when she gets all red and wobbly and gaspy, and fings started to look very soggy in the nose and hiballs department.

Sometimes, Dad says, a man just has to get out of the kitchen before fings get any more well, wet. And fortunately, he did remember to take me with him cos he did remember that leaving Worzel Wooface with a wobbly Mum is very Not. On. and a very long way beyond Stuff. I. Can. Cope. With. In quite actual fact, me and Dad are about as useless as each other at dealing with Soggy Mums ...

But we still did not find Gipsy.

April 22

Mum got up at stoopid o'clock this morning to stand in the kitchen and mutter please-come-home-Gipsy words. And do crying. I aren't hexactly sure how either of these fings are supposed to help get Gipsy to come home but, by the time me and Dad did decide to get up, the kitchen was quite actual sparkly clean, and there was fings in the dishwasher that have not binned used by anyboddedy since I has lived here ... and I don't fink they ever will be.

Dad doesn't know wot to say to Mum to help. So he did say halmost actual nuffink apart from see-you-tonight as he lefted for work. Then he must have forted about it some more as he drove to work, and later on he sended Mum a message asking if there was any noos and sorry-I'm-not-much-help.

But there has binned no noos. And Dad is very luffly to remember not to be a Artless somefing-beginning-with-B, but now Mum is crying again and I are quite actual sure me and Dad did very agree that Mum being art-broken was his actual job and not mine. So I has decided to hide until he comes home.

I would like to say that I are one of those dogs wot does sitting quietly beside his Mum, not-eating-the-tissues-giving-comforting-licks like the dogs in the soppy films on the telly, but I are not. I like fings calm and normal and simple, and if they is not calm and normal and simple I are staying out of the way until they are.

April 25

We is all trying to come to terms with the fact that we will never see Gipsy again. Gipsy has dunned wot we always feared she would, and she's gone to a quiet place where noboddedy can find to make her journey to the Rainbow Bridge. We're all really struggling with the not-knowing and the hoping- she'll-come-home-just-one-more-time so we can say goodbye. And we're stucked in-between wanting to accept she has died and not quite knowing for actual sure. But then Gipsy-the-Cat never did anyfing for anyboddedy's convenience, and maybe we have just got to accept that now she never will.

April 28

Today, Mum and I went for a long walk in all sorts of places a dog really, wheely should not be very hexpected to be on a lead. We went all round the edges of fields and through hedges and down ditches to see if we could find Gipsy's boddedy.

For THE QUITE ^{very} actual LOVE of Worzel Wooface

Mum has accepted that Gipsy has died, but if one more person tells her that cats-often-crawl-off-to-die, I fink she might actual hexplode. She Knows. Cats. Crawl. Off. To Die. but that doesn't make it any actual blinking easier when you is a control-freek who would like to say goodbye to your cat. And also not have to worry and wonder if her boddedy is lying out in the open and not covered up and protected.

So, our massive fuge walk round all of Gipsy's usual places to be was to make sure Mum could not see her. And that I could not find her neither. Cos if neither of us could find her, then she was hopefully safe and protected, and in a special and careful place.

Once I got actual used to the hole not being allowed off the lead and having to stay with Mum, we did have a fabumazing time shoving ourselves into all kinds of small spaces, pretendering to be Gipsy on her hadventures. By the time we got home, me and Mum had twigs all tangled up in our hair, and Mum had a fuge wet patch on her bum where she slidded down a ditch.

But now Mum says she feels betterer: she didn't get to say goodbye to Gipsy or do choosing a place to bury her, but it seems Gipsy has managed to find a perfick spot to do burying herself like the hindependent cat she always was.

And somehow, that isn't such a bad place to start making memberries of Gipsy.

Visit Hubble and Hattie on the web:
www.hubbleandhattie.com • www.hubbleandhattie.blogspot.co.uk • Details of all books
• Special offers • Newsletter • New book news

May

May 1

Dad has dunned a job for Mum and she is really, wheely pleased with it. And Dad. Dad says he has definitely earned a Brownie Point, and please-can-he-go-and-play-on-his-boat-now?

The Brownie Point job is a mirror in the garden to make it look biggerer and posh and booful. It's all quite very actual hartistic, happarently, and just like wot the clever gardeners do do on the telly, and it will be fabumazing once all the plants grow round it with their ickle tendrils. Mum has put the mirror behind a pot of roses so it doesn't look like it is a way through the fence, cos she doesn't want any birds flying into it and getting a Ned Ache or worse.

I don't fink Mum needs to worry too much. That bit of the garden is where all the cats snooze in the sunshine now it is getting warm a-very-nuff to do that fing, and honly a really, wheely stoopid bird would fly three feet from the ground, fight through a prickly rose, and still hexpect to be alive a-very-nuff to bump into the mirror.

The honly fing with an Art-Beat struggling with the hole hidea of the mirror currently is the person who did decide to put it there: Mum. She keeps forgetting about it and catching flashes of herself when she walks past. And having a art attack finking a robber-dobber is creeping up behind her.

May 2

The fuge ginger boyman wants to know if Mum will help with his Diss-the-Nation by reading it and finding all the spellering mistakes. Mum has said of-course and it's-the-very-least-I-can-do after the laptop Confuser disaster.

Tonight she did her bestest best to read it but hardly any of the sentences make sense. Dad reckons Mum is still finking about Gipsy, and that being sad and hupset can make it difficult to concentrate on complercated stuff, and he is probababbly quite actual right. But it might also have somefing to do with the fact that the fuge ginger boyman is in the final year of his bio-chemistry degree, and it is mostly beyond Mum. It is all very complercated and full of science stuff that she doesn't have a blinking clue about.

Heventually, Mum remembered her trick for pretendering to be cleverer than she actual is, but it is a secret so don't tell the fuge ginger boyman. Or Dad. She has dunned gettering away with it for years ... and it's all about bananas. Whenever Dad has got somefing he wants to tell Mum about that is complercated stuff to do with the boat, and she does want to be actual hinterested but not get a three-hour Talk On Hairy-Dymanics, she changes all the complercated science words to banana in her head. It kinda, sorta works, and at least she knows when to nod in the right place.

So, she did the same actual fing with the fuge ginger boyman's Diss-the-Nation, and found some sentences that didn't have a-very-nuff words in them, HAND some spellering mistakes. The fuge ginger boyman was quite actual himpressed with this, and said fings like did-you-really-hunderstand-it, and I-fort-it-might-be-a-bit-too-specialised. Fings did nearly actual end in disaster when he began getting all fusey-tastic about it and asking Mum about the little details, but Mum's binned living with Dad for years and years now, and she wasn't going to fall for that ol' trick. Dad's heven convinced Mum hunderstands the off-side rule in football. She doesn't, but she knows hexactly in wot order to say all the words ... and none of them is banana.

May 4

Dad has gotted a noo job! He is still staying at being foreman at the boat yard but he has binned asked to write for a maggy-zine called *Practical Boat Owner*, and Mum is very, very actual proud of him. But also she keeps laughing a lot and wondering if the Head-Hitter realises that Dad doesn't know how sentences work.

But Dad says they is hinterested in his hexperience and his boat-building, not his sentence-building, so tonight we is celly-brating with wine.

May 5

Someboddedy has binned pooing on the rose that is around Mum's noo mirror. It wasn't actual me, though. I would have to do hacrobatics to hachieve pooing on that actual rose. And also have somefing very quite wrong with me: I don't poo white paint splatters. The best I can do is black paint platters when I has over-hindulged on liver.

Some of the poo has also dunned splattering all over the mirror as-very-well, so now Mum is hupset in case the white paint is a poor birdy's last will and testy-ment where it crashed into the glass.

This morning, Mum has been hanging out of the previously ginger one's bedroom window to see if she can see wot is happening, and if the birdy is still alive. And I has been keepering her comp-knee by hanging out on the previously ginger one's bed. The previously ginger one would prefer it if we did hanging out somewhere else: it is not heven 8 o'clock, and she is definitely not alive yet, and could we all bog off and make her a cuppatea?

May 6

It has binned decided that I will not be having any hinjections this year. Usually I do have somefing called a Booster, wot is a top-up for all the hinjections I had as an ickle baby boykin, but there is some noo finking going on about this, apparently, and it turns out that these Boosters are not quite very actual necessary. In very fact, there is some finking that the boosters could also be doing more-arm-than-good, and I has quite very likely had all the boosts I do need.

I will not be going Habroad this year, and I did have a rabies hinjection last year, anyway, which lasts three years. Rabies hinjections are The Law if I does do going habroad so we can't do skipping that one.

Mum finks it is himportant to make sure Dad is very hinvolved with himportant decisions about me, whether he likes it or not. She tried to do telling Dad all about the hinjection forts, and also about somefing called titre testing, but as she can't work out how to say the word she was not that convincing. I are still not sure if it is said teeter or tighter or heven if both of those are wrong but I are very living with it and so is Dad. Hespecially when Mum did tell him that Sally-the-Vet was doing a talk all about it if he did want to find out actual more.

Dad doesn't very hoften look terry-fried but the fret of a two-hour Talk with Pictures about dog hinjections was a-very-nuff to do it: he has delly-gated all dog hinjection decisions to Mum; he is going to put the sails back on his boat.

May 8

We sawed the poo-ing birdy today and he is not dead. He is a Blackbird.

Today, it is my himportant work to do hadmitting I mighta binned wrong, and say actual sorry to all the Fesants I has dunned hoffending by calling them stoopid. Fesants are not the fickest birds in the world, I has discovered; Blackbirds are. Well, the one in my garden is ... He finks that his reflection in the mirror is another bird, and he has binned pickering a fight with himself all day. And being actual quite noisy and flappy. And splattery.

His most actual lethal weapon for attacking himself is his white poo. You would fink he would realise after a while that he can't drop poo on himself, but he is trying actual quite hard to do this, and Mum's mirror doesn't look booful a-very-tall any more. The previously ginger one reckons it looks hartistic in a very modern kind of way if you like that kind of fing.

Dad doesn't: he says that it looks like a Jackson Pollock, wot I fink is a rude way of saying a right blinking mess ...

May 10

I met a kitten tonight! She is Kite's noo ickle sister, and now that she has settled in, Worzel Wooface was actual allowed to have hintroductions. I was a luffly gentle boykin, and although I was quite actual hinterested to do sniffing and pleased-to-meet-yous, I did remember my manners.

I has metted kittens before. In very fact, a few years ago I did help to look after a litter of kittens wot happened accidentally to a stray cat my famberly founded. But it has binned a long actual time since I has been close up to a kitten, so everyfing was dunned carefully and safely. I are pleased to say that I did remember everyfing and did exerlent lying down and making myself hinteresting-but-not-too-fuge.

The kitten, who is called Nelly, did hexactly wot she wanted. I aren't sure she knewed wot that was cos she can't reliababbly stay sitting up and lookering in different directions without falling over. But I did not hinterfere with the hole falling over fing and pretended I did not see it. Kittens become Cats and Cats is Heasily Hoffended.

And the last fing I do need is a Cat being Hoffended with me. I fink that would hurt.

For THE QUITE very actual LOVE of Worzel Wooface

May 11

Mum's decided to leave the Jackson Pollocking all over the mirror. She hopes it will stop the Blackbird seeing himself and prevent any more rounds of the one-sided Blackbird war. Either that or the Blackbird will give up and realise that all that's happening is that he is getting a quite very actual sore beak from peckering at the glass, wot is a lot actual harderer than the boddedy of another Blackbird. Dad says if he wasn't fick before, he is now, and he must have a Ned hinjury from the bonkers-crazy-peckering. All the tapping and banging makes Dad fink someboddedy is breaking into his shed. He reckons Gandhi will Solve. The. Problem. soon, and it is all about Sir Vival of The Fittest. From the look on Mum's face, I do not fink Mum likes this Sir Vival chap, but Dad says he isn't moving the blinking mirror, he's going down to the boat, and taking his power drill and his Brownie Points with him: he didn't hinvent the Feery of Heevy Lution, Darwin did.

Later on, when the fuge ginger boyman phoned, he hinvented a hole noo version of Top Trumps trying to work out who would win in a game between Darwin and Gandhi – the quite very actual Darwin and Gandhi, not the cat and the Blackbird. Mum says she doesn't want to fink about it, and shouldn't the fuge ginger one be revising for his hexams and we are not naming the Blackbird Darwin: it's going to be horribibble a-very-nuff when Gandhi gets him without that!

May 12
****FUGE NOOS****

Harry is coming to visit next week. His Dad is going to come and do some photo-taking in Southwold at the harbour and at the beach, and we are going to look after Harry whilst he is doing that fing. Cos trying to take photos with a Harry being actual quite hexcited and jogging his arm by tugging on his lead will make the photos all blurry!

I are very actual lookering forward to seeing Harry again. I are wondering if he does remember me, and also if he has remembered that he's not allowed to bark at the cats. Or me. And I are wondering if Mum has remembered that Harry can get onto the kitchen table ...

THE NOO COLLAR

I've got a noo collar
It's very quite bright
I hope that my Dad
Finks it's alright

Mum says it is booful
She loves all the swirls
But Dad might try saying
That flowers are for girls

Well, I fink he's wrong
I look dapper and dashing

And Dad doesn't know
Anyfink about fashion

May 14

Yesterday was a FUGE and Very Himportant Day in my life according to Mum. And anyboddedy who does luff me or is hinterested in doggies. Yesterday, I did decide to play tuggy. For the first time ever, HEVER. Hunfortunately, I did decide to play tuggy with a toy wot Kite had dropped in somefing disgustering. I aren't sure if it was the disgustering stuff stucked on it, or the way that Kite's Mum was holding the toy, but I did run up to her and I did yank and pull and then do some tail wagging and more yanking, and after a while, I did actual realise it was a game and that playing tuggy with a hooman is very actual to be hencouraged.

I aren't hexactly very sure WHY it is such a himportant game. But from the squealing and photoing and the don't-you-dare-let-go and I-don't-care-if-you've-got-poo-on-your-hands words that got said at Kite's Mum, this is hobviously somefing that I are very allowed to actual do. And if I do decide to do it again, I will be tolded I are a good boy, and there will be happy dances from Mum.

Me and Kite's Mum did decide we can do coping with the happy dances and the squeaking and the other redickerless fings Mum did do after my tuggy game, although it was actual quite loud and hembarrassing. But we is both very Living. With. It. seeing as it made Mum so happy. Kite's Mum is also much actual happier now she has dunned washering her hands ...

May 15

I has just had a hunexpected and too-close-hencounter with a small person's pink scooter wot had binned lefted lying on the ground in a hunexpected place. Fortunately, there was no small person actual attached to the scooter, so the honly-fing-with-an-art-beat wot got scared and hoffended was me. And Mum. She had to have a hunexpected and too-close hencounter with a cuppatea. I fink this is fair. She says I are a blithering hidiot plonker.

May 16

Today, it is all-hands-on-deck, wot is a sailing termy-nology that means everyboddedy has to help. Hunfortunately for Dad, sailing words quite very often are nuffink to do with sailing, and the fuge purple tent and a lot and a lot of boxes full of hinformation leaflets and signs meant that it wasn't sailing we were going to be doing.

Our himportant work today was to be at a fair in Southwold to tell peoples all about the previously ginger one's Warrior Walk that will be happening in Lowestoft in July. The fair was horganised by the Lions, and I are pleased to say that they do have two legs, rather than four, and some of them have big full-of-dinner bellies. None of them did look like they was hungry. They was all very smiley and friendly, and not scary a-very-tall.

After Dad had put up the tent, he did suddenly get an hurgent and himportant phone call asking him if he could possibibbly go into work, wot was actual Not. Helpful. but had to be putted up with. Usually, when Dad has to go to work on a Saturday he does a lot and a lot of sighing, but he did not do too much moaning about it this time, and when he did drive off, he did look more pleased that he should have actual dunned. I do fink it was because of the purple tent, wot is hofficially called a Ga-zeee-bow. Wot it was called was quite actual Not. Himportant. as far as I are concerned: like Dad, I did not want to have anyfing a-very-tall to do with it.

I does not know why hoomans fink Ga-zeee-bows are nice places to sit cos there is all sorts of very fuge dangers hassociated with them. I can't actual specy-fry wot those fuge dangers actual are, but for two hole hours there was a Danger in our Ga-zeee-bow, and I did decide it was my himportant and hessential work to Stop. Mum. Going. In. The. Tent. I did a fabumazing job at this. I has got exerlent brakes for doing stopping when I really, wheely need to do that fing.

After three hours, a quite very actual lot of sighing from Mum, a fuge load of I-really-wheely-need-a-wee messages from the previously ginger one who was stucked with the Danger in the purple tent on her own, some quite actual frantic failed phone calls to Dad to see if he could come back and do taking Worzel Wooface home, fings were starting to look quite very jiggly for the previously ginger one, so I did decide the Danger might have actual gone. To be quite actual sure, I did sneak up round the back of the Ga-zeee-bow and cunningly enter it in a hunexpected way. I fink this did shock and surprise the Danger so much that it had to do runnering away because, once I was in the Ga-zeee-bow, the Danger seemed to have actual disappeared! There was just lots of treats, my comfy bed, a lot and a lot and a LOT of fuge sighs of relief from Mum, swiftly followed by the previously ginger one running as fast as her quite actual short and not very fit legs would carry her to the loo.

When she camed back, the previously ginger one fort about saying some rude words about Worzel Wooface being redickerless, and why-can't-we-have-a-normal-dog? But then Mum reminded her that in a few hours one of them would have to do taking me for a wee and a wander, so perhaps it might be a good hidea to be cheerful and offer me some more bits of cheese or we'll-have-to-go-through-that-palava-again.

At about four o'clock Dad came back and asked whether we had had a nice day and if it had binned a success, and you-all-look-very-comfy. Mum did suggest that she was going to do somefing Not. Comfy. a-very-tall with Dad's phone so he did quickerly hunt around hunder his car seats until he founded it, mainly I do fink so Mum could not do wot she kept saying she wanted to do with it. Then he did very quite wisely decide to do taking me for a nice long walk as far as possibibble from the purple Ga-zeee-bow and the even more purple-in-the-face Mum.

May 17

It's called trigger-stacking and it's the reason I would not go in the Ga-zeee-bow, Mum reckons. It's not: it's called a Ga-zeee-bow and it was purple and flappy and different and new, and that's why I wouldn't go in it. But Mum's looked it up on Uff the Confuser, and even though I can't really remember the hincident with the pink scooter in my brain, my boddedy still can, happarently. So when I had a shock and a hupset with the scooter, all my Oar-moans got over-hexcited and made me want to run away. And they hadn't stopped whizzing round my boddedy when I hen-countered the Ga-zeee-bow, so I just added some more Oar-moans on top of the pink scooter ones, and that's why I did have hystericals and refuse to go in the purple tent.

And really, given all those run-away oar-mones wot were whizzing around my boddedy causing troubles, it is fabumazing that I did heventually have a go at being in the Ga-zeee-bow. Which means that I did actual quite well and Dad is not to call me a plonker or a hidiot. *Those* actual words are reserved for the person who can't be trusted with a mobile phone, Mum says.

May 19

We has not seen Darwin the Blackbird in the garden for a few days, and he has not binned biffing himself in the mirror a-very-tall. We does know this fing because there is no noo poo. You would fink that no noo poo would please Mum, but instead she is sad and worried that Gandhi has dunned catching Darwin, and that it is all her fault for wanting to have a pretty garden.

Mum can't decide if Gandhi should be in the Dog House or if she should be, but today is Gandhi's birfday which seems to have gotted him a get-out-of-the-Dog-House-free card. And a big packet of yummy treats from the previously ginger one.

May 20

Harry is here! He hasn't forgotted me or my famberly, or that if he jumps on the kitchen table he can pinch all the cat food. He has not growed in his boddedy but he does seem larger in other ways. He is very actual happy and conferdent about himself, so he does seem like he is a biggerer boykin.

After Harry had dunned reminding himself how to walk on the lammy-nate floor we had an exerlent playtime of roaring up and down the stairs and dragging the duvet off the bed. Hunfortunately, Harry had forgotted all about Don't Shout at the Cats, so as soon as he started barking, they did all decide to scarper through the cat flap. But Harry will only be with us for one night so they is not going to suffer too actual much.

Everyfing was halmost very quite perfick but tonight I did hear that Hattie-Hazel Princess Poopants was down at the harbour today getting her photo tooked. Mum got to say hello-again as-very-well when I. Was. Not. There, so I did not get to see her. It was quite actual special for Mum to see two of her foster dogs together, and to see their famberlies doing talking together. Mum said I possibibbly, probababbly, halmost definitely would have loved it but it would have been chaos and a lot of lead-tangling as there was No. Way, all three of us coulda binned let off together because there was cows on the marsh, and honly a short bitta beach available to doggies now it is May. Well, it's about half-a-mile long this short bitta beach, but three growed up sighthounds would have roared to the end of that in about 15 seconds, and then kept on going and dunned crashing into sandcastles and small peoples and their Ice-Screams. We would have binned very quite actual bad hambassadors for sighthounds.

May 21

Whilst Harry's Dad was here he did offer to take some pictures of the previously ginger one for the Warrior Walk so we would have somefing to send to the noospapers, and he did also say that he would get the photo of me jumpering over Kite's path that Mum has binned trying to take for the past year and failing big-time-badly.

All Mum's hattempts have either been hinteresting pictures of grass when I has jumped out of the picture before she has pressed the button, or a messy ginger smudge where I has been going too actual fast for Mum's camera to cope with.

I are sorry to say that I has not been able to be at all very helpful with this fing. Path-flying is complercated a-very-nuff without me having to worry about whether Mum is ready to press the button on her camera. It's full speed or not at all. I are not a birdy with wings who can do floating in the air – I is a fuge Lurcher, usually with a Kite-pretending-to-be-fierce coming up behind me, so stopping in mid-air is not a hoption.

Harry's Dad said he would do getting the photo for Mum, and when he arrived at Kite's garden he brought the most ignormous camera I has ever seen. It had a fuge trumpet fing on the front of it that all us doggies did fink needed to be hinspected to make sure it wasn't going to make any strange noises, hespecially as Harry's Dad did do lying down on the ground to make sure he had the right angle. After that, Kite did decide it was her himportant work to be Harry's Dad's nable assistant and sat on his head, so Harry had to stand in as my zoomie pacemaker. And we did do getting the photo in one click! Honly one click, heven with Kite being probababbly-not-that-very-actual-helpful, and Mum laughing and Kite's Mum panicking about wot Harry's Dad might be lying in ...

May 22

Darwin is quite very actual alive! And well! And making a squawky racket at the top of the napple tree. Gandhi did have a go at climbing up to see him, but got distracted when he heard Dad opening the larder door, and suddenly found himself on the end of a too-thin-bitta-tree, and then heven more suddenly on the Not-Grass-Really-Carpet facilerty with a clatter and a fump.

It was all very actual hexciting, hespecially as I forgot for a small moment about Not. Chasing. The Cats. I fink Dad must have binned not concentrating and finking sailing forts because he was standing on the back doorstep drinking his cuppatea when I did this small bit of chasing, and there were no don't-chase-Gandhi words a-very-tall, just a lot and a lot of gazing into the distance. I aren't completely sure Mum would actual happrove, but Dad did not want to do discussing it when I came back into the kitchen: he just began to sing a weird song about old women and shoes ...

Darwin-the-Blackbird seems to have losted all hinterest in attacking himself in the mirror. Either that or he has forgotted about it. It's called Ham-Knees-Ear. Banging your head on glass can do that to you, Dad says.

Most very himportantly, Mum is feeling very actual relieved and has dunned taking herself out of the Dog House, and Gandhi, who was never in there due to his birfday not-in-the-Dog House card, has been gived lots of strokes and you-wouldn't-kill-Darwin-would-you? Feeling relieved seems to make some of the peoples in this house very quite Oppy-Mistic I do fink. And deluded, Dad reckons, and if Darwin has got any sense a-very-tall, he will do choosering somewhere else to live whilst he still can ...

May 24

Dad says that Darwin has found out about girls, and then him and Mum had a fuge great discussion about whether you could call a female Blackbird a girl and whether that was sexist, cos she would be a woman, not a girl, hespecially if she is finking of making baby birdies with Darwin. And is also possibibbly, probababbly even ficker than Darwin himself if you do finking about it ... Darwin can't help being fick – he's kinda stucked with himself – but the future Mrs Darwin, as everyboddedy is now calling her, could choose any blinking Blackbird in the hole wide world, and she'd have to be even actual ficker than Darwin if she choosed him.

Dad is now wondering whether he isn't nearly as clever as he once fort he was ...

May 25

Noboddedy has actual very seen Mrs Darwin yet. Mum reckons Darwin hasn't managed to find anyboddedy, being that he's so fick and all the female Blackbirds has made better choices and found a mate who doesn't attack himself in a mirror, or choose a garden to build a nest with four cats and a Lurcher living in it. He isn't hexactly a great prospect and, hopefully, all the Feery of Heevy Lution stuff will stop him having baby birdies, and then there won't be any fick little Darwins making a mess of Mum's mirror. Or hanging around making a fuge noise trying to get eated by Gandhi.

May 26

Today, Darwin has decided that sitting in a tree yellering about I-am-a-fabumazing-Blackbird-come-and-make-baby-birdies-with-me isn't gettering him anywhere and he Must. Try. Harder.

It would very actual seem that Darwin is possibibbly one of those jeany-us types: he is actual rubbish at halmost everyfing but he has got a talent. It turns out that Darwin can Walk and Talk (or Fly and Squawk which is heven more himpressive) at the same time without bumping into anyfing, and that is more than the fuge ginger boyman can do, heven when he has not drinked too much Cider.

It's a Man Fing, Mum reckons. As soon as the fuge ginger boyman gets all hinterested in talking, he wanders around all over the path and bumps into Mum or whoever he is talking to. It drives everyboddedy bonkers, and if I are out on a walk when he starts talking, I do find that I need to watch where he is putting his feet because he ... doesn't.

I aren't completely very sure if it is a Man Fing or a fuge ginger boyman fing because the honly other man I do a lot and a lot of walking with is Dad. And Dad actual quite rarely says anyfing when we are out on a walk. I used to fink that it was because there wasn't any gaps for him to do talking into, cos Mum filled them all up. But now I are not so very actual sure. Maybe Dad doesn't say anyfink cos he'll fall over or crash into a lamppost if he does ...

Hanyway, Darwin has binned showing off his flying and squawking skills all over the garden today, skimming over the fences and generally showing everyboddedy his jeany-us talent. Heventually, I did decide it was my himportant work to do joining in with my own actual version of the walking/talking/flying/squawking wot is woofing and chasing. I would like to say it did go actual quite well, and at first it did. I did woofing and chasing and managed to keep up with Darwin, and bounced along quite happily all over Mum's plants.

Then I did discover that pots and tubs could be used as wobbly stepping stones to help me keep up with him for heven longerer ... until I found myself taking a giant step into, well, nuffink and no amount of walking or talking or squawking or flying was going to stop me crashing down: I are quite very actual certain I did give Mum heven more hevidence about Darwin being a jeany-us and Men can't Walk and Talk. It was not helegant or digger-ni-fried. And there was No. Sympathy. or poor Worzel words a-very-tall. Just a lot of noise about my booful flowers! And stop-chasing-blinking-Darwin.

May 28

There *is* a Mrs Darwin. Mum saw her today pulling all the loose stuff off the edges of her hanging baskets so she can build a nest. Mum hasn't seen Darwin doing this fing yet and happarently he's leaving it all to her, and that is typical-and-hardly-surprising, and when is Dad going to fix up the wobbly stick fing she wants on the wall in the garden?

May 30

According to Mum, our house looks like a bom has hit it. I has never actual hexperienced a bom but I are quite actual sure that boms don't forget to stack the dishwasher and get behind with the washering, which is wot our house currently does look like. But there is a Very Himportant Person from a radio station coming round tomorrow to talk to the previously ginger one about the Warrior Walk, and Mum has decided that they can do their talking in the garden. Dad reckons it's so she doesn't have to do the voovering.

May 31

According to the fuge ginger boyman, Mum and Dad really, wheely need to stop-with-the-casual-sexism. I has got No Hidea wot casual sexism is but it is somefing to do with Mums doing voovering and Dads putting up sticks. The fuge ginger boyman was doing actual quite well at convincing Mum and Dad that they Must. Do. Betterer. but then he spoiled it all by tripping up a kerb and standing on my paw, and also saying that female Blackbirds make nests, not male Blackbirds, and it's not Darwin's job so he isn't hinterfering.

Later on, when Mum asked Dad if he was ever going to do the wobbly stick, he started to mutter about not-my-job-not-hinterfering. It didn't go down very actual well: Dad is not a Blackbird, Mum says, and she's not cooking tea until it's done.

June

June 1

I has a noo job! It is quite very actual hexciting and I will be fabumazing at it, Mum says. A couple of days a week I are going to be a Barney Hentertainer, wot does sound quite actual himportant and perfick for a luffly boykin. Barney's Mum has dunned going back to work and Barney cannot go with her. So, it is going to be my himportant work to go and see Barney in the middle of the day and do playing with him. And making sure he does a wee and that kind of fing.

Today, I did go round to Barney's house to check out his facilerties and to do making sure that Barney did like me. Barney's facilerties are quite actual betterer than mine so I are pleased I will be being a Barney Hentertainer at his house rather than at mine. He has a big garden, and bestest of-very-all he has got a pond! And not just *any* pond but one wot is soon to be a Nex-Pond and get filled in and there's-no-fish-in-it-the-Heron-saw-to-that. So I has binned tolded it is actual okay for me to do jumpering in it. In very general, I do fink this is a Good Fing because, to be actual honest, I does not fink I would be able to do resisting jumpering in the pond after a run around, and if I had to do that fing, then the playtime with Barney would be hawful and boring and hinvolve a lot and a lot of Not-in-the-pond-Worzel words and be very quite frustratering.

It is going to be my himportant work to do making sure Barney gets some hexercise. But not too much, and I are not to be hoffended if he does just do some small runs and then decides he has had a-very-nuff. In fact, it is himportant that I does not do too very actual much hinsistering because, hunfortunately, Barney does seem to have binned born without a nose. Well, he has got the black bit that goes on the end of his nose but the bit between the black bit and his hiballs is missing. I has got No Hidea where it has gonned but, happarently, he is meant to look like that. And it's cute. And dorable.

Heventually, I do fink there will be nuffink else in this world lefted to surprise me about hoomans but not yet, hobviously. Barney might look cute and dorable but I aren't sure if that quite very actual makes up for the fact that he can't get a-very-nuff air into his boddedy so he can do runnering around and playing and normal, everyday fings that doggies should be able to do. Mum says that Barney isn't as bad as some doggies she has seen, but I do find that actual hard to believe. She says at least Barney can do some runnering and chasing a ball before he clapses in a heap, but it will be our himportant work to make sure he doesn't overdo it. And we mustn't let Barney get too actual hot, neither.

I fink it will be very okay though: I are actual quite good at belting-around-in-circles on my own and being hentertaining for peoples who sit on their bums and do nuffink and claim to be Worzelling, so I are quite very sure I can do the same fing for Barney. I just aren't sure that Barney wants to sit and do nuffink. He's honly two years old, and I are sure he'd rather be racing around with me, not sittering on the grass in the shade cos he can't breathe.

continued page 73

Snow is confuddling huntil it is fabumazing!

Playing on the beach with Teddy-Samoyeds.

With kite on the marsh ...

... she likes to blow bubbles in water. I does not like that fing.

The beach in winter is fabumazing but windy.

Jumpering over beach fuff.

I killed Mum's hat which caused hystericals.

Helping with SARDOG (Southwold & Reydon Dog Owners Group) hinformation.

Playing with my oldest friend, Merlin.

Finding yourself on a poster is sometimes Not. Helpful.

Me and my tracker.

Did you call me?

Runnening up hills is easier 4br dogs than it is for hoomans.

Mum and I has gotta bond, which is why I has not done bogging off ...

Fat Frank not quite actual breakering the scales.

Mouse being checked over by Sally the Vet.

Our last photos of Gipsy.

June 2

Mum is trying to work out where Mrs Darwin has builded her nest, and also trying to work out if she should try to discourage the birdies from building it in our garden if she finds it before they finish. She says it's a die-lemma. On the one hand she knows that disturbing nesting birds is actual very quite wrong, but on the other hand, she knows that if they build it in our garden, it will possibibbly, probababbly, halmost definitely end in disaster.

Dad reckons Mum should just let the Blackbirds make their own decisions, and perhaps Mum might like to fink about not controlling every single blinking fing that goes on around here, and concentrate on more himportant fings. Like helping him find his ketchup. The Blackbirds managed to look after themselves perfickly well before Mum tooked an hinterest in them, and she should let Nature take care of itself.

June 3

The fuge ginger boyman phoned last night quite very actual hupset. Noboddedy knowed the right words to say but Mum and Dad did give it their best shot ... All of the fuge ginger boyman's hexams had gonned okay until now, and he did get a very quite actual high mark for his Diss-the-Nation. But his hexam today was horribibble and he could hardly do any writing words on it a-very-tall. He doesn't fink he will fail his degree but he knows he isn't going to get the mark that he did hope for. All he can fink of is that he is nearly £60,000 in debt, and it all feels like it has all binned for nuffink.

Mum tried to tell him that all of the fings he has done whilst he was at Universally – like first aid and chopping down trees and singing in Hoppy-ras – will probababbly end up being far more actual himportant than the mark he does get for his degree, and that noboddedy will care in a couple of years' time, anyway, but that isn't helping actual much right very now.

Dad told him to go and drink Cider.

I fink the fuge ginger one will mainly be following Dad's hadvices ...

HOLES

Someboddedy dugged a big hole
It wasn't actual me
I are hentirely hinnocent
As you can actual see

Someboddedy made a fuge mess
Diggering the very big hole
They was quite very actual hunting
A sneaky-deaky mole

Someboddedy dugged a big hole
But wot is very weird
I seem to have a lot of mud
Stickering to my beard

For THE QUITE ~very~ actual LOVE of Worzel Wooface

Someboddedy's in the Dog House
For diggering in the wrong place
That someboddedy, it turns out
Is me Worzel. Woo. Face

June 5

I went round to see Barney today and it did go actual quite well. At first, Barney was not sure he did want to come outside but, after a while, and when he did realise that nuffink scary or worrying was going to happen, he did come out for a wee. And then we did a little bit of runnering round his pond, and Mum did do hencouraging him to play with a ball.

Barney and me does have the same hidea about balls. We like the hole hidea of them being waved in the air by hoomans, and we love all the talk about 'watch it' and 'are you ready?' And then we do do exerlent puttering our bodies in the 'we is ready, frow it hooman' position. Now, if that was the hole point of ball frowing me and Barney would get a Brownie Point ... but it isn't. And that's where it does go a bit very actual wrong.

Barney and I did both go leapering after the ball, and we sawed it and we founded it. And then we did both do completely actual very hignoring it. Neither of us was hinterested in the hole picking-it-up or the bringing-it-back bit. So then we did runnering back to Mum, waiting to do the bit we is both good at which is lookering-hexcited-and-hinterested.

According to Mum we is both blinking useless, and go-and-pick-it-up! We has both decided that Mum is much, much actual talented and betterer at the go-and-pick-it-up bit. So she can do that part and we will do the everyfing else. It is called the Vision of Labour, and I do fink we has gotted it just about actual right.

June 7

It's all very well and good Dad saying do-you-remember-where-I-left-my-ketchup and let-Nature-take-care-of-itself, but Nature has not dunned a-very-nuff remembering about Gandhi. Gandhi is prowlering round the garden like a cat on a murderous mission looking for Darwin. At least he was ... until I becamed a dog with a mission and hencouraged him to do his prowlering somewhere else. Now he is doing all his Darwin-hunting about six inches away from where he was before but on the other side of the fence where I can't reach him. Mum has decided the other side of the fence is a bit of the world she can't be responsibibble for: this side of the fence is Our Garden, and inside it it is her himportant work to make sure that Gandhi does not catch Darwin. The other side of the fence is the Rest of the World, where Nature is very In. Charge.

I aren't sure Darwin or even Gandhi can tell the difference between the two sides of the fence to be quite actual honest, and they has probababbly not noticed who is In. Charge. in each part anyway. Fences seem to be fings to sit on as far as cats and birds are actual concerned. They don't get in the way like they do for dogs, so the hole point of them is probababbly losted on them. Dad reckons if it's keeping Mum happy finking that death will not become Darwin in her garden then, as far as he is concerned, a Hapocalips can happen

74

on the other side of the fence so long as someboddedy buys some more ketchup tomorrow ...

June 8

Today, it was much, much hotterer than it was the last time I wented to see Barney, and he was not keen a-very-tall on runnering about. I did exerlent hentertaining him by whizzing round and round. I tried to biff him with my nose to see if that would actual hencourage him to come and join in, but he did look most actual hunimpressed. I did quickly deciding that it was a Bad Hidea and then entertaining him whizzing round and round. and then exerlent hignoring him, and then I did jump in the pond for the first time. I had a fabumazing time lummoxing about. Gettering out of the pond was actual quite easy once I did discover that there was some steppy ledges hunder the mud and the water. They were quite actual slippery with mud and green slidy sludge, so I had to have a couple of goes at it. Mum did make one of her halarmed faces but I did a lot and a lot of reassuring her that I was quite very actual in my helement. There is honly one bit of the pond where it is too deep for me to stand up, and finding it was not helegant and did halmost result in my head going Hunder. The. Water. wot is against everyfing I do believe in. After that, I did decide it was time to get out ... but then I gotted in again because ponds is hirresistibibble. So long as you know where the deep bit is.

Whenever I got out of the pond, Mum was not that keen to do saying hello-luffly-boykin. She said that I did smell-like-a-drain and could I go-away-and-shake-somewhere else. But mainly not in her coffee.

****FORTS ON SWIMMING****

- I are still not actual completely very sure if I can swim
- The honly way to find out if I can swim would be to do it, and so far in my life I has dunned exerlent havoiding of it
- Finding out might mean I have a bit of a disaster, wot would count as A. Bad. Day.
- All the hoomans in my famberly can swim: the fuge ginger boyman started off the worst at it but he is now the best. He did have a big problem with sinking but Percy Veering helped until he was fabumazing at it
- The previously ginger one did find it very actual easy, so Percy Veering did not have to help her
- The previously ginger one feels the same way as me about putting her head Hunder. The. Water. I aren't too actual worried about messing up my hair, though
- Mum is a good swimmer cos she is clumsy, and spended a lot and a lot of time falling off boats when she was ickle so she had proper swimming lessons
- Dad says he is an okay swimmer, but he is a sailor and falling off boats generally makes them go slower so he doesn't do it. And could Mum. Not. Either. It wouldn't make the boat go slower and he might have to go back and get her
- Okay – he would **HAVE** to go back and get her and could she stop looking at him like that
- Mum wants me to have a go at hydro-ferapy to find out if I can really, wheely swim. But I does not have poorly legs and she is not sure the hydro-ferapy peoples want their pools used by dogs that have got nuffink wrong with them. Apart from a control-freeky Mum

For THE QUITE very actual LOVE of Worzel Wooface

June 10

I are not wot you would call a Hambitious Boykin. I does not want to win prizes or learn to do clever tricks or climb mountains or anyfing like that, so my list of Fings to Actual Hachieve is quite very short. In quite very fact, there has honly ever binned one fing on my list:

CATCH. THE. MOLE.

And I are quite actual pleased to say that my life's work is dunned, cos yesterday I did do that fing. Wot is more than can be saided for my Mum and Kite's Dad, who is useless and rubbish and ... I fink I'd better start at the beginning ...

I was minding my own actual business and having a sniff and a smell around the edge of Kite's garden when The Mole popped up his head. There was no scrabbling or hexcitable Worzel Wooface hinvolved cos that would have meant Mum Coming. To. Hinvestigate ... so The Mole just felled into my mouth. You will be pleased to know that as well as being reasonababbly cat-safe, I has also discovered I are mole-safe and did not do chomping on The Mole. But I did do exerlent runnering to the bottom of the garden with him to hinvestigate my fuge success. And then he wriggled. And squiggled. So I dropped him. And then he Shouted. At. Me.

I do not know if you has ever hearded a mole doing shoutering but it is an Orrendous and Terry-frying noise. Wot should be hobserved from a safe distance. Hunfortunately, the shoutering mole did attract some attention from Mum, so he decided it was time to actual Hexit Stage North and began diggering to try and bury himself. You has got No. Hidea. how quickerly a mole can do burrowing away when he can hear a Mum fundering up the grass coming to hinvestigate.

By the time she gotted there The Mole had halmost disappeared under the ground again. But this was not hacceptababbly. Noboddedy did want to hurt The Mole (even Kite's Dad who was called to come and help, and appeared armed with a trowel wot did cause Mum to have some don't-you-dare-kill-it hystericals). But everyboddedy did want to heject him from the garden; firstly, and very himportantly, to stop him diggering up more of the grass, and secondly to stop Worzel diggering up the grass to try and see his moley friend. The trowel was useless at diggering up The Mole, and did more harm than actual good to the grass, so in the end Mum gotted hold of The Mole by the scruff of the neck and plopped him into the field, where he has got about 70 hacres to dig up. Kite's Mum is not feeling Oppy-Mistic. He'll Be Back, she reckons ...

June 11

Mum says it's a good blinking job that there is nuffink actual alive in Barney's pond, and that it is going to get actual filled in, cos today when we did go round to see him, the level of the water was lower. Much lower. And she doesn't fink it can be put down to Heevap Or Nation. Heevap Or Nation is wot happens when the sun sucks all the water into the clouds so they can do raining, but if the sun had sucked as much water out of the pond as it did look like, then we would get rained on for hever. She finks there is a Hole in the Pond, and it's all my fault and hobviously I do need my nails trimming. When I hearded these forts, I did wot any sensibibble dog would do and hidded my feet. In the Pond.

The water in the pond is now a fabumazing colour: pooey brown with a touch of green: the kind of colour that if you founded it in a baby's nappy, you would be callering the doctor and having hystericals. It smells about as bad as it looks as well, Mum says.

In very general, me and Mum does have very different finking forts about good smells and bad smells. Mainly, everyfing she says does smell wonderful I does fink is boring, and the fings I fink smell fabumazing she would rather she did not have to smell a-very-tall. Hespecially like Barney's pond, when the smells are all over me.

Fortunately, Barney did fink I smelled actual quite hinteresting, and did decide to follow me all round the garden with his nose up my bum trying to find out as much as he could do about the bottom of the pond. Barney is not actual keen on swimmering so I do fink this is the honly way he can do discovering wot the pond smells like.

Mum says she is quite very glad that Barney didn't jump in the pond. First-of-very-all, Barney isn't actual her dog and second-of-very-all, Barney is quite small, and fird-of-very-all, Barney isn't very fit. And if he did jumpering in there and looked like he was actual struggling with swimmering or trying to get out, she might have to Get Hinvolved. And Getting Hinvolved might actual mean having to Get In. Somefing she does Not. Want. To. Do. Hespecially as the mud and the green slimy stuff is now starting to make bubbles and pop like a swamp from an Orror film.

June 13

I has founded Mrs Darwin's nest, and if you did fink I was casting nasturtiums at Darwin and Mrs Darwin for being fick, then I has now got heven more hevidence. They has builded the nest in the complercated hedge near the napple tree, which is about the most stoopid place you could build a nest HEVER. Well, building it in our garden was an actual stoopid place to start with, but they has builded their nest honly five feet off the ground! And there is at least another four feet of hedge above it that would have binned a lot actual safer, I do fink. Fortunately, the nest is buried actual quite deep in the hedge, wot we are all hoping will be too much heffort for Gandhi to get to, hespecially when there is a fuge bag of kibble in the larder and a bunch of hoomans who don't seem to be able to communercate about have-you-fed-the-cats-today?

June 15

Before I does say anyfing I want to make it blinking clear that jumpering in the pond was Not. My. Hidea. That is not to actual say that I didn't fink it was a fabumazing hidea: I did. But I was gived permission. Worzel Wooface was tolded that he could jump in the pond and swim about, and also that the pond was going to be filled in and ...

When we wented round to let out Barney today, I did exerlent hignoring

For THE QUITE ˅very actual LOVE of Worzel Wooface

of all the door opening and come-on-Barney-up-you-get hencouraging talk, and also Barney-shift-you-must-need-a-wee hinsistent hinstructions and headed for the pond. And jumped in.

It's not really a pond any more. It's more of a muddy puddle about five feet wide and four feet long, and the water honly comes up to my elbows when I are standing up. Apart from the deep bit. But I did remember where that hole to the other side of the planet was and did lying down in the shallow bit instead ... wot caused some hinteresting bubbly pops. Quite very actual quickly, I changed from being a ginger-coloured doggy to a sewage-coloured swamp monster, which was quite very smelly-but-to-be-hexpected. And I was quite very actual Minding. My. Own. Business and finking luffly smelly forts about mud and gunge and how much I could take home with me when Mum shrieked, and there were a lot and a very fuge lot of loud words about Worzel-gerrout-the-pond-gerrout-GERROUT!

After all this hurgent gerrout stuff I did actual hexpect us to be belting off at a billion miles an hour out of the garden, down the drive, down the road, into our house, up the stairs and hiding hunder the duvet. It was that kind of shriek. So you can himagine my surprise when I did bravely runnering away, got to the gate and discovered that Mum was not behind me. Hinstead, she was bended over the pond Frozen To The Spot and staring into the water.

It is a very strange fact that when someboddedy is doing somefing quite actual hintently in a concentrated sort of way, it is himpossibibble to not do that fing as well. So I did joining in with the staring and gazing. And then Barney got up and staggered over to the pond to see wot all the fuss was about, and soon all of us were standing Frozed-like, gazing into the pond. One of us knew wot we was doing but to be quite actual truthful, it wasn't me or Barney. And quite not normally for actual Mum, No Words were coming out of her mouth to give me or Barney any kinda clue.

And then ... Somefing. Moved. And so did Mum. Well, her mouth did. Twice in very fact. One of the words she did say is not repeatable in plite comp-knee. The other one was a very quite Armless word wot did not deserve the other one that she joined onto it a-very-tall: Fish.

June 16

The fuge ginger boyman, who seems to have becomed Mum's Blackbird hexpert, says Darwin and Mrs Darwin should by now be on at least their second lot of babies, so they must either be rubbish at the hole being-a-Blackbird-fing or all their previous hattempts have gonned horribibbly wrong.

Then he wented on to say that Blackbirds build their nests in the low down bit of a hedge so that when the babies are ready to leave the nest, they can plop down onto the ground without breakering themselves. Happarently, Blackbirds take ages to learn to fly, and so the baby ones hide in the hundergrowth until their wings grow, and they squeak and flap about so that their parents carry on feeding them on the ground. In our garden. With four cats and a Lurcher.

Mum says she needs wine ...

June 18

Mum says she feels hawful now. She would never have letted me in the pond if she'd known it was home to another creature. And now she's got a moral hobligation to the little lonely fish, and she needs more moral hobligations like a hole in the head but it is now our himportant work to get the fish out of the pond before all the water drains out. And then find a pond wot hasn't got a hole in it and isn't going to have a blinking-great-oaf-of-a-Lurcher jumpering about in it.

Dad says it will be an easy job to get a fish out of the pond now that it's got no water in it, and Mum should just get on with and he's going to work. And not coming fishing. Not even this evening. Not even if it is supposed to be feraputic, and it shouldn't-take-more-than-one-person-to-catch-a-fish.

June 19

Dad wants to know wot Mum plans to do with the fish when she has caughted it. Cos we don't have a pond and he has got no hintention of diggering one. And no, he doesn't need another hobby – fish-keepering is complercated and hexpensive, but most-of-very-all, we don't need Any. More. Pets. And can she stop calling the fish Fred like it's already part of the famberly ...

June 20

It is quite very actual not fair and not reasonababble I do fink for hoomans to have two fings wot they call beds, and actual hexpect me to hunderstand that I are allowed on one of them and not on the other blinking one. Mum says I are allowed on the beds in the house but I aren't allowed on the beds in the garden. This should be completely actual hunderstandable, happarently, and how-many-times-do-I-have-to-tell-you?

Lots of times, it turns out. Hespecially when Dad is a bad hinfluence and leaves the front door open, and finks it is a bit, actual quite a lot and a lot funny. And takes photos cos Worzel-likes-your-newly-made bed-and-he's-just-helping-isn't-he-sweet?

Mum says I are not helping and Dad is not funny. And he is also in the Dog House. I, on the other actual paw, are hiding on the hindoor, hupstairs bed, and I shall do coming downstairs when the hoomans in this famberly have Resolved. Their. Differences. And also comed up with another word for bed cos one is very quite actual not enuff ...

June 22

It's honly a month to go now until the Warrior Walk, and there is so much actual stuff here it is getting very quite hard for a luffly boykin to find somewhere to lie down that isn't either in the actual way or where it's likely I will get stuff dropped on my blinking head.

As well as doing the actual walk and also tellering the hole world to have different forts about peoples with a Men-Tall Elf, there is also some fundraising happening as-very-well. Happarently, there is going to be face-painting and a raffle, and that flipping Gaz-eee-bow is going to be coming out again. I has decided that I will be doing the walk but I aren't having nuffink to do with that purple flappy tent. Dad says he is going to be very In. Charge. of me for that day, and he will do the walk as well but mainly he keeps muttering thank-you-thank-you-thank-you in my hearholes.

I don't fink Dad likes the Gaz-eee-bow neither but he is a growed-up hooman who can't do actual hadmitting he is scared of it, so he is letting me do all the I-aren't-going-in-there-hever-again stuff for the both of us.

June 24

Fishing is supposed to be a calm, relaxing activity. Sittering by the bank of a river finking gentle forts and planning wot to have for tea. Maybe having a snooze and letting the breeze waft over your boddedy. That kind of fing. And mainly not catching anyfing but gettering out of the house and not being told to do the voovering. Feraputic fishing is mainly done by Dads, I do fink.

Today, we has not been doing that kind of feraputic fishing. We has binned doing frustrated fishing wot is a completely different kind of activity, and in very general, I does not recommend it. A-very-tall. And it is not ferauptic unless that do mean that peoples will need ferapy *afterwards*.

Most of today we has dunned sittering by the pond looking at it. I are no longer allowed to jump in the pond: the pond is no longer my playground, it is an Abbey Tat. And an Abbey Tat wot is shrinking big-time-badly. And also himpossibibble to see anyfing in cos the water is all muddy and slimy.

Every so often the mud goes pop, tricking Mum into dippering her net into the water, swooping it hunderneath and stirring up the water. And, most himportantly, Missing The Fish. I do fink that any fish wot has managed to survive the Heron, the water going down, HAND a Lurcher leapering all over it, isn't going to sit at the top of the pond doing wiggly semaphore for 'catch me.'

Mum says she is not going to give up on Fred the Fish, though. And we will be back tomorrow.

Dad says the honly way Mum is going to catch Fred is to empty more of the water from the pond. Honly Dad didn't call him Fred cos Dad is not getting sucked into becoming hemotionally attached to a fish who isn't staying and won't be going into a pond that Dad isn't going to dig!

June 25

We has called in Hinternational Rescue of the fish kind. And they is FAB. I aren't completely sure if they is Hinternational Fish Rescue, to be very actual honest, but a man and his daughter who do have hexperience of catchering fish camed over and helped Mum empty some of the water out of the pond so they could definitely catch Fred and, most himportantly, they did say that once they caughted Fred he could go and live in their ignormous pond with their other fish.

The daughter part of Hinternational Fish Rescue had long boots up to her bum so she did very actual get in the pond and managed not to fall down the hole to the other side of the planet. And she caughted Fred the Fish quite actual quickly. I would like to say that after Fred was caught, Mum did a happy dance, they all had a cuppatea, and everyboddedy – but most hespecially Fred the Fish – did live happily ever after. But that would not be a true fing ... cos after the daughter caughted Fred the Fish, she caughted another Fred the Fish. And another one. And then some more, and it turns out all those pops and bubbles wasn't mud and ...

Now we has got 67 flipping Fred the Fishes living in every bucket and if-it-holds-water-it-will-have-to-do pot that we own. Dad is due home from work in half-an-hour, and Mum keeps singing a song about sixty ways to leave your lover.

June 27

I are very quite actual pleased to say that Dad has not dunned leaving his lover. Or the 67 fish. Or me, most himportantly. It turns out Dad knows about fish, and that they are a type of fish called Carp, which is how they did manage to survive in that gloopy water because they used to live in rice fields with hardly any water.

And then he spended some more time in the garden lookering at the fish and talking to them and making sure they had some food. But they is still all over our garden in pots and fings, and it is quite actual hard to find somewhere for a luffly boykin to do a wee.

June 29

Gran has got a pond! I did not know this fing. Happarently, it is round the corner in a bit of her garden that I are not allowed to visit cos it is full of a greenhouse, and also sometimes Roxy, who is Gran's not-quite-right-in-the-head Bearded Collie who I do pretending does not hexist: it's safest that way.

Today, we will be taking six of the Fred-Fish to live in her pond. Dad did show Mum how to collect them in a bag, and make sure there was plenty of air in the bag, and then did more wittling on about being quick to get them into proper water and not putting too many in a bag and not stressing them out ...

On the one hand, Mum is quite actual pleased that Dad is so hinterested and concerned about the fish. On the other hand, she is beginning to wonder if the key to getting an hattentive and concerned husband is to float around in a bucket being orange and not saying anyfing.

June 30

Gran does not have a pond. She has a hole in the ground wot used to be a pond. Discovering when you are holding a plastic bag full of gasping fish isn't helpful. A-very-tall. Mum began to have a bit of a panic after all Dad's hinstructions about not letting them be in the bag for too actual long, and did wot she quite actual often does and got loud and hembarrassing. And knocked on all Gran's neighbours doors until she found someboddedy who was in and who had a pond.

Heventually, Mum camed back with an empty bag, and Gran was very actual concerned about who Mum had given the fish to, and I-hope-you-haven't-upset-my-neighbours. Now, I aren't the smartest doggy in the hole wide

For THE QUITE *very* actual LOVE of Worzel Wooface

world but I are, possibibbly, maybe, probababbly the most actual sensy-tive.
So I are pleased to say that I was safely at home, so did not have to witness the
fuge deep breath and long hexplosion of words-beginning-with-B and fish and
hidiots and ponds and not ponds and all-your-neighbours-think-you're-nuts-
anyway words that came out of Mum.

People's say having fish is relaxing and good for getting rid of stress.
This, I has decided, is Not. True: it is the Most. Stressful. Fing. I has ever dunned
being hinvolved in.

JULY

July 1

We've seen honly Darwin about for the past week or so, and all of us is actual hoping this doesn't mean that Mrs Darwin has met with an hunfortunate end of the Gandhi kind. Mum is mainly hoping that Mrs Darwin has decided that having babies in our garden with Darwin as the Dad is a redickerless hidea, and has runned off with a more sensibibble Blackbird. Today, though, we saw Darwin disappearing into the hedge with a worm in his mouth, which happarently means that she is probababbly in there making babies.

I aren't hexactly sure how baby birds do get borned, but haccording to the hoomans who do live here, it has somefing to do with Neggs, wot they seem to fink is normal and to be hexpected. I aren't actual sure if it is normal or if my famberly are so confuddled and stressed after all the dramaticals about the fish that they has gived up on trying to keep up with normal and not normal, and have just gotted to the stage where anyfing is fine so long as it doesn't hinvolve them having to fink or do anyfing about it.

But if it is actual true that Darwin is making babies with Neggs, I don't see how it is going to end actual well. Darwin is not the kind of creature that should be very trusted with Neggs: it wasn't that actual long ago he was having hystericals pecking at himself in a mirror and trying to poo on himself. Neggs break if you heven so much as look at them funny, or swipe a box of them off a table with your tail, like Frank did do once.

July 2

I are now quite actual convinced that my hentire famberly has gonned bonkers and completely losted the plot. Happarently, Mrs Darwin is sittering on her Neggs and she can't get off or they will die, which is why Darwin is bringing her food. If I did sittering on a Negg it would be a very ex-Negg quite actual quickerly, so I cannot do hunderstanding why everyboddedy here is thinking that this as a Good Fing ...

July 3

I has been finding out more about Neggs today, and also about Nature, wot is supposed to be The Management and In Charge of everyfing that happens outside. The previously ginger one says that if a birdy sits on a Negg, it stops being a Negg and grows into a baby bird, and then, when it hasn't got anymore room in the Negg, it bashes on the hinside of the shell to crack it and the baby bird comes out! She did it at school, she says – learning how birds get borned; not being in a Negg and fighting her way out. I started to fink that maybe she had got in a muddle with her medicines but Dad says this is quite actual true. And he doesn't take any medicines to get muddled up with. Honly cider, and he hasn't drunked any of that tonight, so it must be true. It's all part of Heevy Lution, apparently.

For THE QUITE ^very ^actual LOVE of Worzel Wooface

Nature is quite very actual odd and strange, I has decided. And Heevy Lution is heven more bonkers-crazy, and I does want nuffink to actual do with it. Fings might be weird and odd and strange in my house HAND my garden, but I are quite actual glad there is a fence and a Mum, and *she* is In Charge of wot actual happens hinside it. Nature, I has decided, can have the rest of the blinking world: I'm staying in here.

July 4

I are quite actual pleased to say that all the fish have binned founded homes. Apart from four wot seem to have binned forgotted about. I do find it quite actual strange that the fishes wot got forgotted are the ones wot ended up with names ... Fred the Fish is still here, and also Fick Fish wot jumped out of his container and managed to get gived Cee-Pee-Ahh by the previously ginger one. And also Fat Fish and Frilly Fish.

We also seem to have a noo fuge barrel with plants and a bubbly fountain in it. Wot also is not getting talked about. The previously ginger one says it is a Hunspoken Hagreement: we must say nuffink and pretend it has always binned there.

July 5

We're going to Corny Wall this weekend! It is all very actual hunexpected but it needs to be dunned, Dad says. We will not be staying with Granny Wendy and Granddad like we usually do, but in a nice hotel near to where they do live.

Granny Wendy isn't well and we are going to go to say goodbye to her. Hoomans, I has discovered, sometimes know when they is going to die, and they get actual quite horganised and make sure that they see all the himportant peoples in their lives before that happens.

Some people do believe that dogs know when they is going to die, and perhaps they do, but not several months before it is going to happen, and they don't generally horganise a barbeque either. Mum says it will be my himportant work to be a luffly gentle boykin and also to be an exerlent distraction: everyboddedy will know hexactly why they are visiting Granny Wendy but they won't want to talk about it, and having me about will give them somefing else to fink and smile about.

I like Granny Wendy and Granny Wendy finks I are fabumazing, which has surprised her a lot and a lot because. in very general. she isn't a Dog Person. Well, you wouldn't be either if you was chased all the way to work by a fuge, cross and scary dog when you was seventeen on your way to your first ever day at work. And ripped your tights and ended up flustered and crying and hupset. I fink it would put you off for very actual life, too.

But Granny Wendy and I has reached a hunderstanding: I will be gentle and she tells me everyfing she is doing and where she is going to put her feet before she does put them there. And then she lets me decide if I would like to do moving out of the way. People who do give me some hoptions are my very bestest kind of people. Granny Wendy doesn't fink she is a Dog Person but she really, wheely is.

July 7

It's Mum and Dad's 10th wedding hanniversary today. A wedding hanniversary is where two peoples do remember the time that they did say I do to each other. Mum says the 'I do' is all very well and good, and there is fuge bits of 'do' that Mum likes about being married to Dad. But there is quite a few bits of being married to Dad wot Mum would like to put into a fuge box and lock it. She says she probababbly shoulda got a pre-numpty hagreement, but that would not have worked cos she did not know about some of these fings before they got married, and even if she had, she didn't know how much they would very acctual hannoy her after ten blinking years.

****FORTS ON ALL THE FINGS MUM SAYS 'I DON'T' TO DAD****

- Letting your luffly boykin Worzel Wooface into the front garden. Cos I are completely very actual determined to do diggering up her plants. And she does not happreciate it
- Taking off 17 layers of clothes at once and dumping them all on the floor so she has to play undoing them all before they can go in the washing machine
- Refusering to take any of us pets to the vet in case they might do dying. Wot is not brave and not being a responsibibble hadult
- Trying to end harguments by saying that Mum is hunaware or by saying she is 'hard work' when he has No Hidea how actual quite aware she is, and how much harderer work she could be if she wanted to ...
- Playing loud Confuser games without his headphones on so the voices of fifteen strange men fill up her house. And also their rude words
- Playing loud Confuser games with his headphones on so that he is completely deaf and doesn't hear Mum calling him.
- Tellering Mum quite actual proudly that he has dunned some hoovering For. Her
- Somehow convincing everyboddedy in her famberly that he is a habsolute saint for puttering up with Mum
- Snoring
- Sneezing more than three times in a row

Hoomans are strange, I do fink. Cos apart from the snoring, wot gets discussed halmost every blinking week, she hardly ever, hever talks about these other fings. And the snoring is the One. Fing. that Dad probababbly can't do anyfing about. And all the other fings wot he could do somefing about she just stomps around the house muttering about to Worzel Wooface when he isn't actual here. Maybe she doesn't want the actual harguments. Or, more very actual quite likely, she doesn't want to know all the fings that Dad didn't say I Do, to, too.

July 8

I heard Mum talking to the hedge today. Nuffink surprises me any actual more. She mutters at the seeds in her front garden all the time, and you don't heven want to know what she calls the snails before she hurls them out of the garden and back into nature's bit of the world.

Today's hedge-talking was actual directed at Mrs Darwin, but because she can't see Mrs Darwin buried in the hedge with her nest and her Neggs, she had to have a bit of a guess where the nest actual is. She's completely actual wrong – it's about four feet further along in the hedge than where she finks it is – but then she doesn't have my sniffing and smelling skills.

So the bit of the hedge where Mum finks Mrs Darwin is hidded has binned hinformed that Mum is going away for four days, and can she please keep her babies hinside their Neggs until Mum gets back because Gandhi is going to be hunsupervised, and she can't be responsibibble for what he does whilst she is away. Gandhi has been told he is to stay out of the hedge, and Kite's Mum has binned asked to make sure the cats always have plenty of food.

Dad says it is our himportant work to just-smile-and-nod-Worzel when Mum gets like this. And also make sure that the dustpan and brush is in an hobvious place when we leave so that Kite's Mum can clear up any hevidence ...

July 9

We is on our way to Corny Wall, and so very far the trip has been actual quite okay. Currently, we is near somewhere called Hex-hitter, which is where my friends, Dawn and Toby, do live. Fings were going to be fabumazing because they does have a noo Lurcher puppy called Jack, and it was quite actual planned that we would do having a playtime and a trip to the pub so we could get to know each actual other. But when we were travelling down we did hear that Jack had binned poorly sick, and so he couldn't come out to play after-very-all, and he would have to stay hindoors and not meet Worzel Wooface.

I did decide it was my himportant and hessential work to make up for this disappointment for Jack by weeing all over his favourite field, so now it does smell a lot and a lot of Other Dog. I do fink it is now a far more actual hinteresting field than the one he did have before, and I does hope he happreciates my hard work. And I has got to say it was blinking hard work cos like all fields in this part of Ingerland, it had been tipped up on its end. And although I could do charging to the bottom in about six seconds, it did take everyboddedy forever to clamber back up to the top. Hespecially Jack's Mum, Dawn, who was wearing flip-flops.

You might remember that I are not keen on flip-flops but these ones were actual okay because they were useless at flipping and flopping and hinstead decided to be slip-slops. And they did slip-slop off Dawn's feet and then actual help her get to the bottom of the steep grassy slope a lot and a lot faster than she did actual intend. After that, Mum and Dawn did decide to stay down there and mutter at Dawn's seeds and shout about snails. Even though nature might have tipped itself sideways in the West Country the snails do still cause troubles, and also understand hexactly the same words-beginning-with-B that the ones in Suffolk do!

I hope Jack will quite very henjoy his field when he is allowed to go and play in it again. I fink I has dunned compensatering him for missing his trip out, and I are hoping he doesn't feel cheated and hoffended that his Mum and Dad did meet another dog and go to the pub. Jack is not even a year old, yet, so he doesn't know that this is Very. Not. On. His Mum and Dad might do gettering away with it ...

July 10

Today, I has binned to see Huncle Keith at his boat, and I did also have a hunexpected flying lesson. Huncle Keith's boat is not in the water. Dad doesn't fink Huncle Keith's boat will ever live in the water, and although it is teck-nick-ally a boat, cos it is boat-shaped and possibibbly, probabably might float if it very actual had to, it is more of a boat-shaped caravan. So, anyboddedy who wants to visit Huncle Keith's boat has to climb up some steps. I has dunned this fing lots and lots of times before, and I are quite actual good at it, which is more than can be said for Mum. She is not keen on steps or ladders or anyfing that isn't the ground. She wobbles and panics standing on a chair at home, though she is very used to puttering on a brave face in front of me. And Dad.

Today, it was all going quite actual well, and Mum made it to the top without having an art attack, and it's-hardly-Mount-Everest-is-it-you-daft-woman, when I did have a sudden change of art about visiting Huncle Keith's boat. So, I jumped off the steps sideways and nearly tooked Mum with me cos she was still holding onto my lead. Somefing I had not taked into consideration. A-very-tall. And hinstead of finding my feet safely back on the ground, I did briefly and very quite halarming find I had becomed the world's first doggy Trapped Peas Artist dangling from my harness.

Heventually, Mum remembered wot the word LEGGO! actual means. It tooked far actual longer than I would have actual preferred, and Dad had to shout it at her several times before she did do this fing so I could do dropping to the ground. Fortunately, my hunexpected and not very quite hagreeable go at being the world's first Trapped Peas Dog was so actual confuddling that all I could do when my feet hit the ground was stand there. And give Mum A. Look. By rights, I shoulda bogged off big-time-badly but when a luffly boykin has just dunned dangling in the air wondering how he gotted into this mess, and more very himportantly, how he will get out of it, sensibibble forts do not always come to mind. So I just stooded there, until Dad could squeeze past Mum having hystericals at the top of the steps, jump down the steps as nimbly as a fifty-something-year-old-man actual can, scoop me up and carry me onto the boat. Any plans, Dad reckons, he might have had for his knees in the next six weeks have now been cancelled.

July 11

It was Granny Wendy's barbecue today. I fink it did go quite actual well from the hoomans' point of view. We had the barbecue at Auntie Sue and Uncle Dave's, where I did get to do some Himportant Work chasing huninvited Hinvading Cats from the garden, and exerlent hignoring of the cats wot owned the place. Some peoples do fink that luffly boykins cannot tell the difference between these two types of actual cats, but we can. Well, I can! General Frank of the Ginger Militia did teach me all about it when I was an ickle Worzel Wooface. A cat-wot-owns-the-place does scowling and sittering still and creates a space all around him, wot luffly boykins are not allowed to enter. A Hunivited Hinvader Cat does not do this fing. He looks furtive; not sertive. So, I did chase away the furtive-looking ones and left the sertive ones to sit by the barbecue supervising Uncle Dave's steak cookering ...

For THE QUITE ˅very actual LOVE of Worzel Wooface

July 12

I are home from my trip to Corny Wall, and I are quite very actual not pleased to say that I has bought some huninvited and hunwanted Stower Ways home with me from that blinking dog-friendly hotel we did stay at. The owners of the hotel said they had a special room for dogs wot they was quite actual proud about, and they did hobviously fink that all they needed to do was to say 'we is dog-friendly' to actual *be* dog-friendly. However, they did need to do a bit very more than that: like spray the blinking carpet for a start!

Dad spotted them all over my belly on the last night in the hotel, when I did turning myself hupside down for an exerlent snooze on my bed after my busy time eatering sneaky bitsa steak at the barbecue. Fleas! Fundreds of flipping fleas! All over Worzel Wooface! Trying to nibble bitsa me and then discovering that I did taste too much of Billy-No-Mates to be worth a second bite! And then I did have to actual hendure a fuge hinspection from Mum whilst she tried to pick them all off and squish them between her nails. Squishering fleas is a great way to let all your finking forts wash over you, Mum says, and quite very feraputic when you need to do some finking about the lovely evening you has had, but also want to do feeling a bit actual sad at the same time, and can't find the right words or forts or feelings.

Hunless you is the luffly boykin gettering squished that is! After a very while it does get quite boring ... and all my hoptions about want to change position and put my leg in a different place were limited by a determined Mum muttering about keep-still-I've-nearly-got-the-last-one.

Honly she hadn't, of course. As soon as she gotted into bed and I laid down on the carpet again, another gang jumped onto me. So today, instead of having a nice relaxing day hinspecting the garden and reminding the cats that I do still actual hexist, and they has not inherited a small three-bedroom house with a large shed, a fish pond, and a hedge full of takeaway baby Blackbirds, I has spended most of the time runnering away from Mum and her spot-on Flea Drops, wot I don't usually have to very hendure but there-are-too-many-of-them and it's-July and come-here-you-blithering-hidiot.

In the end I did have to have my Lead. Put. On. in the house and got satted on the sofa, where I did exerlent shaking and making it quite very actual clear that I would DIE rather than have a small amount of liquid put on the back of my neck.

July 14

Judgering by the amount of going backwards and forwards in the hedge and the noise coming from it, Mrs Darwin has dunned having her babies. Well, at least one. Or at least one has survived. We aren't sure how many are in the nest, and we will probababbly never actual know because none of us are allowed to do going pokering around to see. Hespecially not Gandhi, who is now being shutted in at night along with all the other cats, whether he actual likes it or not.

'Not' is mainly Gandhi's hopinion, and he has tried several actual times to bash his way out through the catflap, and heven tried to get Frank to put his badger-like weight behind the problem, but it's no good. To get Frank to become the Basher of Barriers, he has to be stucked on the wrong side of the food bowl, and he isn't: he's sat on the kitchen table which is hexactly where he wants to be.

This morning when Mum came down she discovered a dirty protest against cat curfew. In Mum's washing basket. Wot had a nice clean load of washing waiting to be hung out in the morning. Mum finks it was Gandhi, and he has binned called all sorts of Not. Nice. Names. Honly it wasn't Gandhi – it was Mouse, who has never really hunderstood about hinside and outside, ever since some daft hooman bought a tree into the sitting room one Crispmas ...

July 17

Have you seen Mostyn? He has gonned missing in Halesworth, and it is not like him a-very-tall. Mostyn is Gandhi's litter bruvver that you might remember did get borned in this house. And if you does not remember, it is actual okay because he does look hexactly like Gandhi. Honly his eyes aren't quite so actual close together, and he isn't sitting in our napple tree wondering wot time Blackbird Telly will start this morning ... So, if you do see him, please can you let his Mum know because she is going a bit bonkers-crazy worrying about him. Fanking oo kindly.

July 19

I can see Gandhi is going to get ignormous over the next few weeks: he honly has to look at the hedge and Mum rattles a cup of cat food. Fortunately, Gandhi does not have a-very-nuff gentleman bits to do making babies. Otherwise I do fink the hole Feery of Heevy Lution would create a lot and a lot of cats wot do fink they only have to look at hedges for cat kibble to magically appear ...

July 20

The fuge ginger boyman has grab-you-lated! He has got letters after his name, and we is all very actual proud of him! That flipping hexam wot had all the questions he did not very hexpect messed up his marks, though, and he missed gettering the result he did want by a quarter of a purrsent. I has got No Hidea wot a purrsent is, let a-very-lone a quarter of one, but happarently it's an eeny weeny fing. Hunfortunately, he did do a science hexam where the peoples in charge can do maffs, and even though the fuge ginger boyman did ask them to do addering up the numbers again, it did make no difference. Mum finks it is cruel and artless, and I fink she is more crosser than the fuge ginger boyman.

At first, the fuge ginger boyman did not want to be batman-flapping-around-in-a-black-gown, and go to his grab-you-lation celly-bration, because he was so actual disappointed with his result, but Mum has actual hinsisted. She did not get to go to her grab-you-lation cos she was up-the-duff with the fuge ginger boyman. And seeing as she missed hers, she was quite very actual going to his!

Unless he could come up with a biggerer reason for not going than Mum being fuge and waddling and ... After the fuge ginger boyman did realise he couldn't do being up-the-duff, he did decide he would go.

Part of the grab-you-lation did actual hinvolve peoples doing walking in a straight line to collect their himportant ser-tiffy-kit. You would be quite very actual blinking HAMAZED at how many peoples who are clever a-very-nuff to do gettering a Science Degree can't walk in a straight line, or not step on each other's toes, heven when they isn't actual hexpected to do talking at the same time. The previously ginger one reckons Mum might not have fort a-very-nuff about how much fizzy wine was swilling around hinside the peoples doing the marching to collect their ser-tiffy-kits. But I fink she might have a point ...

July 22

At the grab-you-lation, Mum did meet the fuge ginger boyman's noo girlfriend. She says she is tall and slim and very actual booful and hincredibly clever. Mum did fink she was a luffly kind person, and she didn't have no problems walking in a straight line, heven though she was wearing redickerless shoes.

I does know that hoomans do say that you shouldn't judge peoples by first himpressions, but I are a dog, and with dogs first himpressions count for halmost everyfing. If you is sertive or furtive or nervous or scary the first time I do meet you, I will fink that you will be like that every time we meet. Hoomans can do giving peoples a second chance quite actual easily; a lot and a lot easier than dogs can do, anyway. I aren't so sure about this noo girlfriend, though, because I has received hinformation wot I do not actual happrove of, and I does know that Dad will very hagree with me. With Knobs On.

The fuge ginger boyman's noo girlfriend is called Kat. For the last few weeks, Mum has been muttering about us being A Man Down. But she doesn't mean A Man Man, she means a Cat Man. Ever since Gipsy did die, and we has honly had four cats, Mum has been wondering about a kitten. Dad says we need more pets like a Hole In the Head and we've-just-got-four-fish. Dad doesn't do saying No very actual hoften but he has been saying No a-very-lot about another cat.

And now it do seem that we has got another blinkering one whether we like it or actual not!

July 24

Quite actual sometimes, hole HOURS do go by when I do not get to see Mum. When that do happen and then she comes back, Mum is quite very nonchy-lant, and does not make a super big fing about have-oo-missed-me. It is all about seppy-ration anxiety Mum says.

Fortunately, Kite does not suffer from seppy-ration anxiety like Mum hobviously does. When Kite has not seen me for a few days, she is always very actual fusey-tastic about seeing me, and makes a hignormous fuss and tries to sit on my head. I aren't sure how Kite worked out that sitting on my actual head is the bestest way to let me know she has missed me, and is quite very pleased to see me. Perhaps she finks if she sits on my head then the rest of me won't be able to go anywhere without her again.

July 24

I do look like a Pun Crocker, haccording to Dad. I has got No Hidea wot a Pun Crocker is but, happarantely, the Mow-He-Can Kite's Mum gives me when she gives me luffly scritches along my back is all part of being a Pun Crocker.

Tonight, I did finish off my Pun Crocker doggy look by skidding all over the newly-cut grass in Kite's garden, so now I has gotted green legs and feet wot finishes-off-the-look and you-look-like-Wellard-Worz. Hoomans have far too actual many words I do fink and not nearly a-very-nuff of them make any blinkering sense. Mum was honly an ickle girl when Pun Crockers roamed the earth, she says, and she fort they were terry-frying, but Dad was a teenager and he reckons Pun Crockers fort they looked Wellard, when really they were called fings like Norman, and weren't nearly as rufty-nufty as they did look.

I are beginning to feel some haffinities with the Pun Crockers. I might look Wellard with green feet and my Mow-He-Can, but when Kite's Dad put away the grass cutting machine, I did have an Orribibble fright and had to run away bravely to hide at the top of the garden. And I aren't coming back down again until everyboddedy stops talking about safety pins ...

July 26

Mostyn has comed back to his home! Well, he's back home, which isn't the same actual fing because he didn't choose to come home, but had to be collected and broughted back home in a basket in the back of a car. For some reason, he was found about half-a-mile from his actual house, hangering around in a big car park with the fuge lorries wot take parcels and furniture and tins of food up and down the country. One of the lorry driver peoples tooked him to a vet who did scan his microchip, and then his Mum wented to get him.

Mostyn is in the Dog House. Being in the Dog House when you is a cat must be Orrendous, I do fink. But his Mum Does. Not. Care, she says. He did give her a fuge worry and hupset so he is not going to be allowed out again until he has forgotted all about lorries and running away to Scotland. Or at least until his Mum can work out why he was so far from home and trying to go on a hunauthorised holibob. I fink his Mum does care, wotever she actual says. I fink she cares a lot and a lot.

July 29

Tomorrow, all the previously ginger one's hard actual work and being in the noospapers is going to be quite very worth it, Mum says. The previously ginger one says she's not so sure, and maybe everyboddedy can do the Warrior Walk without actual her, and she will stay in bed and hide? Cos she's changed her mind about it all and she doesn't want to be a warrior, and she doesn't want to go for a walk, and most of-very-all, she doesn't want to talk about it.

Dad says Mum should leave the previously ginger one alone and let her get over the panic. She'll be fine, he says. I aren't convinced, but the fuge ginger boyman will be arriving first fing tomorrow and is bringing his noo girlfriend for Dad and me to meet, and if anyboddedy can get the previously ginger one to feel better about everyfing, he can.

And his noo girlfriend being there should be a-very-nuff to get her out of bed because the previously ginger one won't be able to actual resist coming to hinspect her. Mainly cos she is called Kat, I do fink.

For THE QUITE _very_ ~~actual~~ LOVE of Worzel Wooface

July 30 (early)
It's Warrior Walk Day at the Lions Festival in Lowestoft! I are very actual lookering forward to seeing the Lions again later on. And their hotdogs. Mum lefted the house at stoopid o'clock this morning to put up the Gaz-eee-bows with some peoples who will be running the Men-Tall Elf stall, and giving out leaflets and painting butterflies on people's faces. Dad has gone to get the fuge ginger boyman and the Kat Lady from the station and I are home alone. More home alone than I are supposed to be, to be quite actual honest. The previously ginger one has bogged off big-time-badly ...

July 30 (a bit later)
Dad is back with the fuge ginger boyman and the Kat Lady. She does not look like a Kat or smell like a Kat, but Dad seems to like her as much as he do like all cats. So far, he has dunned resisting stroking her, and I have decided it is my himportant work to say hello to her and stop Dad making a complete hidiot of himself, seeing as Mum is not here to do that job.

Fortunately, the fuge ginger boyman has dunned noticing that the previously ginger one is quite actual habsent without leave, and everyboddedy is in hagreement that she must be founded before Mum finds out. Mum is due to meet our Em-Pee and our Em-Ee-Pee in about an hour, and fings could get quite very actual tricky if she finds out whilst she is doing that fing that the previously ginger one has gone Hay-Wol ...

July 30 (a bit more later)
We has founded the previously ginger one! She wasn't that very losted, to be quite actual honest, but she is hiding and isn't coming out. She's too-scared-to-do-the-walk and she-can't-do-it and someone-will-have-to-tell-Mum. There are times when my rubbish habilities to commbefore with hoomans are very actual frustrating. Today is not one of those times, though ...

July 30 (heven more later after three cups of coffee, a lot of talking, and redickerless amounts of make-up)
I has got No Hidea wot the Kat Lady said to the previously ginger one, but it was hobviously magic words. And quite very actual nuffink about Em-Pees and sponsorship monies, or the people who would be turning up to do the walk with her, which is wot Mum would have said and made everyfing worse, the fuge ginger boyman reckons. Most actual very himportantly, we is now all in the car; everyboddedy has remembered to wear their t-shirts, and everyfing is going to be actual okay.

July 31
Dad says I was a fabumazing boykin at the Warrior Walk, and did exerlent keeping him comp-knee at the back so that he could make sure that nobodedy got lefted behind. At the end of the walk there was lots of cheers and clappering, and then everyboddedy wented onto the stage and the polly-tickle people said some words, and Mum did, too, which she wasn't hexpecting to do. Dad didn't go onto the stage because he was with me and I didn't fancy

going up the steps. Hinstead, we wented and stood and listened to Mum doing talking and also watchering some Samba ladies warming up. I aren't surprised they had to do warming up to be quite actual honest. I fink they coulda dunned with wearing a few more jumpers and jeans, or a lot and a lot more fevvers.

Being a warrior isn't about being super-brave all the time. It's about being brave when you don't want to be and you've got a spot on your forehead and you haven't had a-very-nuff sleep. It's about walking along being very quite hordinary so that other people who do feel the same can do the same fing. Like me going into a purple Ga-zeee-bow when you is convinced there is a Danger in there, heven though everyboddedy finks you are being a bit, well, redickerless.

And it wasn't just the previously ginger one who was a Brave Warrior at the walk. A fundred other people came to walk with her, and some of them were big-burly-rufty-tufty-looking men who had to live with their own Men-Tall Elf, and mums with pushchairs, and just a-very-bout every single type of hooman you can himagine. It was a fabumazing success, Mum says. It was habsolutely hexhausting and terry-frying, the previously ginger one reckons, but she did also raise nearly a fousand pounds for Mind, so it was halmost, just-about-worth-it.

And can she have a cuppatea now?

AUGUST

August 1

It has tooked them months and probababbly several goes, and it's a lot and a lot later in the year than it should binned, but Darwin and Mrs Darwin have finally managed to get a baby Darwin to live long a-very-nuff to plop out of the nest and hop around under the hedge. Heavy Lution needs to catch up a bit, I do fink. Heventually, the fuge ginger boyman says, the honly Blackbirds that will survive will be the ones who either live in the middle of nowhere, or ones who decide that hanging around on the ground flapping and squeaking, and generally making themselves look dee-lishous to cats, is a very quite stoopid hidea, and that growing a few more wings before falling out of a nest might be a betterer longer term plan. Trouble is, he reckons, that will take a few fousand actual years. In the meantime, Darwin and Mrs Darwin are going to have to rely on Mum.

And if that's their only hoption, then Baby Darwin really, wheely better hurry up and learn to fly cos Mum and Dad are going on their holibobs in five days' time, and honly the previously ginger one will be here. She has decided that if she is brave a-very-nuff to be a Warrior, then she is perfickly capababble of being Home. Alone. for a fortnight. I aren't convinced: she honly had to be a warrior for about five hours and that's not nearly as hard as being a hadult.

August 2

I fink Baby Darwin must have hearded that Mum is going away in a few days because he has managed to work out how to get out from hunder the hedge and into a tree. That's all he's hachieved, though. He's like a little old man who has finally managed to get up the stairs and sit in a comfy chair, and now he's Not. Moving. for the rest of the day.

Baby Darwin doesn't look like Darwin a-very-tall, but he is just as noisy. He has fuge great yellow lines round his mouth so that Darwin and Mrs Darwin can see where to put the food, and he seems to need a lot and a lot of it. His mouth is like a glow-in-the-dark dustbin, and his Mum and Dad keep shovelling worms into it hendlessly, but he always seems to be hungry and shoutering about it. He does remind me of the fuge ginger boyman a lot and a very lot.

Baby Darwin is not the honly one to have hearded about Mum and Dad's holibob: the fuge ginger boyman and the Kat Lady have decided that they are going, too. Mum finks that it is quite actual fabumazing that the fuge ginger boyman wants to come on holibobs with her and Dad. Hand bring the Kat Lady. It does mean that Mum and Dad are still himportant peoples in their son's life, and he is choosering to spend actual time with them. It means they have been good parents, she reckons.

Dad reckons the fuge ginger boyman has got no monies and is after a cheap holibob, and hoping that Dad is going to buy all the beer ...

August 3

The previously ginger one is still planning on staying Home. Alone. when everyboddedy else is on the boat. I still aren't sure this is a fabumazing hidea. I fink she knows where the freezer is, and I does know she can make herself a cuppatea, but that's about it. Kite's Mum says she is going to feed the cats – which the previously ginger one could probababbly do – but that way, she'll have a good hexcuse to pop in twice a day and make sure that the previously ginger one hasn't blowed herself up trying to do somefing complercated with the Confusers. And also that she is still actual alive to be very actual honest: the previously ginger one still has her Men-Tall Elf, but she is nearly 20, and there is honly so much lookering after and watching and hinsisting that you can do when someboddedy is teck-nick-ally a hadult ,and says they want to stay at home.

August 4

He's dunned it! Or, she's dunned it! One of them fings! Either way, Baby Darwin has managed to fly out of the garden and into the rest of the world. It's called Fledging, happarently. I fink Fledging is another way of saying it's-a-flipping-miracle-now-go-and-be-someboddedy-else's-problem. And stay out of our garden and away from Gandhi, and please, please, PLEASE don't come back. Or turn out to be as fick as your stoopid parents. Mum and Dad are off on their holibobs and nature is now In. Charge. of Baby Darwin.

August 6

Now that Mum and Dad are on their holibobs, I are staying with Gran-the-Dog-Hexpert, and I have to actual say that I are henjoying it. There is a good actual reason for this, but I aren't supposed to be henjoying it for that reason, so it is my himportant work to make somefing up hinstead. Happarently, sometimes you cannot tell the quite very actual truth, and have to be a plite and luffly boykin.

So, I should do tellering you that I are having a great time because the weather is very quite sunny, and Iona-the-fishwife has decided not to shout at me quite so very hoften, and because Gran-the-Dog-Hexpert has letted me have lots of playtimes with Mattie, who is her quite very jolly and cheerful Bearded Collie who finks I are exerlent fun. And all of these fings is true but aren't the real actual reason.

The real reason, I are shamed to say, is because Baillie, Gran's very quite actual bossy and scary other Collie has died. And cos he is dead he isn't here to herd me, and very, very slowly curl his lip at me in a way that most hoomans can't see hunder all his fur. But I used to see it and although Iona's fishwife screaming and banshee doing is frightening in a loud and sudden sort of way, nuffink gived me the wobbles and the shakes like Baillie's silent-but-terry-frying lip curls. It was a-very-nuff to make me not want to do anyfing but try to work out how to be as small and hinvisible as possibibble. Which is quite hard when you is an ignormous ginger Lurcher and all Gran-the Dog-Hexpert's other dogs are Cavalier King Charles Spaniels. You kinda stick out. And Up. Mainly Up, to be quite actual honest ...

For THE QUITE ^{very} actual LOVE of worzel wooface

None of the hoomans did really notice this fing. Everyboddedy had gotted so used to Baillie and is hobsessions with balls and herding fings and bossing and boinging around that they just tooked it for actual granted. And it was my himportant work to actual put up with it because it was Baillie's house and he was a senior doggy wot must be hobeyed. So I did do that fing and tried very actual hard to be umble and not get in his way.

Now that he is not here, it is like the end of a bit of music when all the clashing notes do stop arguing with each other and gets to the chord that everybody has binned waiting for. And they can do clapping and going to have a drink of wine at the pub. All the tension has gone. And I has not had to be worried about being perfick every single second of every day. Heven Iona's banshee fishwife noises aren't bothering me quite so actual much. And tonight, I did feel brave and conferdent a-very-nuff to do Sharing A Bed. This is all noo and probababbly the bravest fing I has ever dunned at Gran-the-Dog-Hexpert's house. It was so brave that Gran did hunt around for hever lookering for a camera, wot was buried about nine feet deep hunder a load of hagility stuff, to Record. My. Hachievement.

Now, Gran reckons, all she has to do is persuade me to eat somefing else other than chicken wings, cos she has decided it is redickerless that I do refuse to eat anyfing else when I are with her and she is Going. To. Solve. It. I has becomed A Project and a Mission and it will be hachieved whether I like it or very not. Baillie might be actual gonned but his spirit lives on ...

August 8

I saw the previously ginger one today! The good noos is that she is still very actual alive. But not well. The hole Home. Alone fing was actual very hentertaining for a few days, but now she has runned out of milk and it's-too-far-to-walk and her bike has got a puncture that she can't fix. She wants to go to the Neverlands and have a cuppatea with Mum. There is a shop wot sells milk a mile away that she could walk to very actual more easily than get to the bit of Habroad where Mum and Dad are staying, and Gran did offer to take her to get some milk, wot would cut out all the walking and possibibbly dying of hexhaustion bits. And she also suggested that the previously ginger one came to stay with me at her house but that isn't wot she wants to do. A-very-tall. And once the previously ginger one has maded up her mind about somefing, it is halmost himpossible for her to be happy until it's done.

Dad fort about phoning the previously ginger to tell her that she had choosed to stay at home so she should quite very put up with it. But I fink he knew that Mum would have hystericals and not henjoy the rest of the holibobs because she'd be worrying. So heven though it is going to cost them two days of their holibobs and about a billion pounds, the previously ginger one is going to get on a train and then a ferry and then another train and then a bus to get to the bit of Habroad in the middle of nowhere that Mum and Dad are currently at. And then have a cuppatea.

August 9

The fuge ginger boyman and the Kat Lady have decided it is their himportant work to go and collect the previously ginger one from the ferry. Wot will actual hinvolve a lot more busses and trains and trams, and even more billions of monies: it is far too complercated for the previously ginger one to manage on her own, they say. And then hopefully, the fuge ginger boyman will remember how to get back to the boat ... that Dad won't have moved because he's so blinking frustrated and hannoyed that his holibob has been completely hijacked by growed-up children who either drink all his Cider, spend all his monies, or make Mum have sleepless nights and drive her to the point of hinsanity.

August 10

The previously ginger one has made it safely to Habroad and the boat and most himportantly to Mum. And Dad who decided that moving the boat was possibibbly, probababbly not the best hidea in the world. And also pointless when it has a GeePeeEss Tracker on it that the fuge ginger boyman hunderstands, and a Mum who would definitely sit on the pontoon and refuse to go with him.

When Gran phoned to checked everyfing was actual okay, she was quite very hinterested to know that the previously ginger one is not the honly extra actual passenger they've picked up neither, and so I did also get to hear all about the Seagull wot did itching a ride with them.

Seagulls are not fick birds like fesants or Blackbirds, and Heevy Lution is hobviously working a lot and a lot faster for them I do fink. Either that or they is hincredibly lazy, but then the fuge ginger boyman told Gran that being lazy is all part of Heevy Lution as well, and any bird wot can find an easier way of doing fings has got more time to make babies. And then those babies will also learn that sittering on a sailing boat rather than flying up the Grevelingenmeer is a lot and a lot quickerer and easier than flying. And also they have redickerlessly hexcited peoples on them wot are more than happy to hand over all the ham and anyfing else wot can go in a sandwich, just so they can get a photo with the ignormous birdy to send to all their friends in Ingerland.

August 11

Gran's Project and Mission to get me to eat somefing else other than chicken wings is all very quite sigh-and-triffic, but currently she is gettering halmost nowhere. Some fings I are very actual happy to eat, like dried liver or tripe. Duck is honly good for rolling in, I has decided, and pizzles are exerlent food when you don't fink too actual hard about them being the gentleman bit of a cow. Well, a bull, to be very actual haccurate. Gran is very actual row-bust about stuff like that, and so is Mum, but I aren't too actual sure how Dad is going to feel about this noo addition to my diet. I do suspect we won't be giving him too much hinformation ... okay, any hinformation.

Anyfing wot I can pick up and turn around with or chomp when I are lying down or having finking forts about is very okay. But anyfing that has to go in a bowl and stay in a bowl, like mince or anyfing sloppy, I do not like, which is a fuge problem when half of my dinners are supposed to be made of that stuff. Gran says she isn't givering up, though, because it's himportant that I get a balanced diet, and she's very fond of me ...

For THE QUITE ~~very~~ actual LOVE of Worzel Wooface

August 13

And Gran *is* actual fond of me! I has never binned too very actual sure about this fing cos Gran doesn't do showering her feelings very much. She is very horganised and fings are always safe and she is the Management of all Managementers, I do fink. But since there is No Baillie here any actual more, and I aren't being such a scaredy boykin this visit who just wants to be very hignored and pretend-I-don't-hexist-fanking-oo-kindly, we has been able to get to know each other some more. I aren't just being very Looked After and Kept Safe, and we has had a very actual good time together. Neither of us is any good at relaxing and just making friends and trusting, so it has tooked a bit of time, but it has binned very actual worth it. Now, she reckons, if I could just eat something-other-than-chicken wings, everyfing would be perfick.

August 15

Gran is still very actual fond of me, heven though a chair wot she did leave me alone with for too actual long hexploded whilst she was out taking Granny Mary to a Opital appointment. I has never had a chair hexplode on me before, but Gran finks it was all her fault and sorry-Worzel-you-must've-been-very-stressed-poor-boy.

August 16

I are a very clever boykin, Gran says. I aren't convinced, to be quite actual honest. All I did was eat the mince, wot was actual quite tasty once I could eat it without half of it getting up my nose and making me feel like I was going to drown.

Feeling like you is drowning is not a good fing when you is trying to eat your breakfast, but that is a quite actual complercated fing for a luffly boykin to try to very hexplain to a hooman when you hasn't got any words, and are relying on them watchering and working it out for themselves. I wasn't heven really sure that was the problem. I just knew I didn't like eatering mince.

But now I do know why I don't like eatering mince and it has got nuffink to do with the taste of it: I just needed a Dog Hexpert like Gran to help me work it out.

August 18

I are back from my holibobs with Gran-the-Dog-Hexpert. When Mum came to pick me up, Gran gave Mum my special tray for eating squishy stuff, and a fuge hexplanation about pointy-nosed dogs not wanting to push their noses into bowls and she-shoulda-fort-of-it-before. And I are to make sure I do carry on being an exerlent boykin about eatering my mince and not do lettering her down.

Gran says I are welcome back Any. Time. She has said this fing before which I do fink was for Mum's benefit and all about making sure that Mum got a holibob. Now, I do fink she really, wheely means it for her as well.

There was not too much talk about the chair wot hexploded when I was lefted alone with it for too long. Gran-the-Dog-Hexpert was very quite frilly-sofical about it but Mum is hoping that it won't happen in our house.

Happarently, once chairs do get the hidea that they can hexplode when you is with them, it is somefing that keeps happening ...

August 19

I aren't sure if Baby Darwin is alive or not because in the two weeks I has binned away, he will have growed up to look like all the other Blackbirds round here, so he could be any one of them.

Happarently, sometimes young Blackbirds do stick around with their Mums and Dads and do helpering them to make another lot of babies, the fuge ginger boyman says. And then he jumped on a train and ranned away back down to Barkshire. Sometimes, Dad reckons, an edercration is Not. Helpful. and next time the fuge ginger boyman mentions Blackbirds he's going to be in the dog house. Mum is hinsisting that Dad gets out the trimmers and makes the hedge a lot and a lot thinner and smaller, and hopes it actual hencourages the Blackbirds to choose somewhere different to make a nest. Like anywhere else in Ingerland and its-not-like-we're-in-the-middle-of-a-city-and-ours-is-the-only-hedge-for-miles.

August 20

Dad is in fuge trouble, I fink. Although I aren't completely actual sure. Yesterday, when Dad did do trimming the hedge, and muttering about the fuge ginger boyman making jobs for him that he didn't need cos he can't stop his mouth from 'edercating' people, he did take his frustrations out on the Napple Tree, wot the fuge ginger boyman still finks of as his, heven though he doesn't really live here anymore, and doesn't do anyfing about the napples or have to actual put up with the wasps or the hornets. But has hystericals if anyboddedy heven finks of chopping it down. *That* napple tree, wot is now a lot smaller than it used to be. Dad's chopped a fuge great lump off it, and then came into the house stomping about and stickering out his chest with a wild-and-scared-and-pleased look on his face wot I has never seen before.

Dee-fi-ants, Mum says it's called. And then she giggled a bit and stroked Dad's arm and smiled and tolded him he was really, wheely naughty though she looked very quite actual pleased. Then they borrowed Kite's Dad's machine that chops up all the twigs and branches to get rid of all the hevidence and he-won't-notice and we-won't-be-telling-him, but mainly a lot of stuff about empty nests. And I don't fink they was talking about the Blackbirds, neither ...

August 22

Dear Mum

Please do not frow tennis balls when I are doing a wee. And then laugh at me when I do carrying on doing that wee whilst chasing after the tennis ball. It is not actual fair for you to do sexist moaning about Dad and the fuge ginger boyman not being able to walk and talk, and then not very happreciate my hefforts at multi-tasking.

From your luffly boykin

Worzel Wooface

Pee-Ess: Kite also says could you not do it. She finks my multi-tasking it actual quite himpressive but she'd rather I didn't widdle on her neck.

For THE QUITE *very* actual LOVE of **worzel wooface**

THE MIDDLE MUM

Mum is not a middleman
And words like 'can you just
get him to?' make her hopping mad
And yell and make a fuss
Mum is not Dad's seckra-tree
She does not want to ask
If he'll fix your printer
The next time he drives past
Mum has a mile-long list of jobs
And fings Dad needs to do
She knows he's very useful
But our house needs him, too
When Mum asks Dad to come and be
Your D-I-Y-ing saviour
Dad somehow finks cos she did ask
That he's done Mum a favour!
Mum doesn't want to moan and nag
It's bad for married life
But all these 'can you just?' requests
Make Mum a boring wife
If you need some help from Dad
Please fink of my Mum's Elf
Don't use her as the middleman
Ask him your blinking self!

August 26

Today is National Dog Day. Happarently, us dogs honly get one day, wot I fink
is a bit very Not. On. and every day should be Dog Day. Every day in my year is
Dog Day but then that is cos I are a dog ... Today, is also Women's Equality Day,
wot is a big complercated fing about why Mum always seems to make more
cuppateas than anyboddedy else in this famberly. Wot Mum would like to shout
at Dad about but he's bogged off down the boat, and also he cooked dinner
last night so that probababbly isn't fair. So, Mum says we will concentrate on it
being National Dog Day and later on we will go and meet Dad after work and
go for a run on the marshes.

August 27

Pip and Merlin have binned staying with us all week wot has been very quite
hexciting! Merlin is a Lurcher. He does not actual look like a Lurcher but his
Mum was a long, tall Lurcher lady so he is one – just the shortest Lurcher that
has even binned born. There has been lots of speccy-lation about what his
Dad might have been. So far we has ruled out Badger and Great Dane, but
everyfing else is still hunder consideration ... Anyway, Pip is a not a Lurcher: she
is a tiny terrorist. And bossy. And now about nine years old wot isn't old but
isn't wot you'd call a whipper-snapper, Dad says.

Today, Pip has decided that the skirting board in our living room is

100

hoffensive or has an art beat. I has never noticed this fing but Pip is very actual wise and sensibibble, but mainly much olderer than me so it has been my himportant work to either actual believe her when she keeps shoutering at it, or hignore her. After quite a bit of trying to believe her, I has decided that it is all too clever for me and I must be Missing. The. Point. So I have decided to hignore her.

Mum says she is also hignoring her. So far, she has been hignoring her for three hours and 46 minutes. She has checked that there is nuffink attached to the skirting board, and she does know it isn't actual alive cos she watched it being drilled to the wall. Fings with art beats generally hobject to being drilled to the wall, Mum says. I aren't making it up: Mum quite very actual talked to the skirting board and Pip and did hexplaining this. I fink hignoring Pip shoutering at the skirting board is causing problems with Mum's brain. Dad reckons we can't blame Pip for all of that, but it-is-starting-to-drive-him-bonkers and stop-it-now-Pip.

August 28
Hoomans do sometimes say that you cannot teach an old dog actual noo tricks. But this is very Not. True. and now I has hevidence. This week the weather has been actual rubbish, and cos Pip can be a bit hindependent about bogging-off-home-without-telling-noboddedy, all our playing has been hindoors or in the garden. Wot is safest and actual avoids that hawkward conversation about did-you-have-a-nice-holibob and sorry-I've-lost-your-dog, which would be a-very-nuff to stop anyboddedy going on holibob ever, hever again. And make them a bit, well, hessy-tent about handing over the nice bottle of French wine they promised to bring you back ...

And whilst we have binned doing hindoor playing, me and Merlin have taughted Pip how to do bitey-facey wot is a Lurcher game that all Lurchers are borned knowing! And Pip is very quite exerlent at it ... now she has hunderstood all the rules ... and also dunned waiting for me to lie down so it is bitey-facey and not bitey-leggy. But the best fing of-very-all, is that Pip can't commentate and do bitey-facey at the same very time. Cos that would be too complercated and like a judo man at the Holympics trying to tell everyboddedy wot is going on at the same time as squashing his ponent from Belly-Roos. And still win a meddle ...

Mum and Dad fink it is fabumazing. And quiet. Finally, actual very halmost liveable levels of noise have binned hachieved. Wot after four years of knowing Pip and puttering up with her hysterical, shrieking commentating is a quite actual relief. And proof that you can teach an old dog noo tricks.

Now Mum's wondering about teachering an old Dad noo tricks, and is looking up courses on Samba Dancing ...

August 29
Dad says he is Not. Doing. Samba. Dancing. He says he doesn't mind doing some more Samba-watching, though. Anyway, Dad says he doesn't need to do Samba Dancing cos he has just proved he is capababble of learning noo tricks and himplementing cunning plans. And more himportantly, he has sorted

For THE QUITE ^very ^actual LOVE of Worzel Wooface

Pip's Complaints to the Management. Pip is very strange, I do fink. She never yells about not being able to get up onto fings cos she is too short. But she has been having complete hystericals All. Day. because she is too tall to get hunder the sofa to reach the squeaky piggy. It must be the honly fing she has ever, hever binned too tall to do. So she has barked, and scrabbled. And runned around in circles. And barked some more and made the fugest Complaints to the Management for hours today. Until Dad decided to prop up Mum's one decent bit of furny-chur with a load of books. It's all very quite stable and safe but it isn't wot you'd call helegant. Or heven practical for sittering on. But it has stopped Pip's hystericals and barkering. And also re-you-nited a hawful lot of odd socks and the shoe-wot-went-missing that Worzel Wooface got hunfairly very quite actual haccused of burying in the garden. It was not buried in the garden but buried hunder the sofa, along with a load of God. Knows. Wot.

So Samba dancing is very off the nagenda for Dad, he says. And I do fink Mum is very quite actual relieved as well. She didn't get to watch the Samba ladies warming up at the Warrior Walk like Dad did. And she didn't realise you had to do wearing quite so many not enuff clothes. Or wiggle actual so very much. Mum says her bum is not sooty-ball for Samba Dancing. A-very-tall.

August 30

This weekend me and Mum are both doing dog-sittering for Gran-the-Dog-Hexpert. This aren't the first time that Mum has dunned dog-sittering for Gran but it is my first time. Usually, I do get to be Dad's Dog when Mum does this fing, but ever since I did staying with Gran for my holibobs, we has all decided I should do more practicing of being at Gran's house whenever I can.

You will be very actual pleased to know that I has not received any hinstructions from Gran-the-Dog-Hexpert about my stay at her house, mainly cos I are a dog and do pay more attention to the other dogs when I are there. Mum has not binned so lucky. Or trusted. Or heven treated like a hadult. She has dunned dog-sittering for Gran quite actual often in the past, but has still received two hole pages of hinstructions and other bits of hinformation, and some of it is actual quite hinsulting. Happarently. And she would like it to be known that she is 47 and has been runnering her own actual house and remembering to shut the fridge door for the past 25 years, in case anyboddedy was unaware of this fing. Like Gran-the-Dog-Hexpert hobviously is ...

August 31

Some of the hinstructions is not so actual hinsulting as it is all about the tablets that the creaking Cavaliers do have to take wot Mum doesn't want to get confuddled about. Don't tell anyboddedy, and hespecially not Gran, but all them blinking Cavs do look the actual same to Mum, and she can't tell which one is which unless she does really, wheely finking about it and comparing them to each actual other. I did fink she should do sniffering them a bit cos that is actual how I tell them apart, but Mum's nose isn't as good as mine and she doesn't seem as fusey-tastic about sniffering dog's bums the way that I are.

And the Cavs don't help a-very-tall. Some of the younger ones don't need any tablets, but they fink that the older ones are gettering extra bits of yumminess – just for being old and special – so they do do lining up in the queue to see if they can get muddled up with a Gerry-Hatrick one and get themselves a treat. I do not do this lining up: I are a cautious boykin and not about to fall for any ol' trick about it's-a-yummy-treat when it has ick-yuck tablets stucked in it. Also, the chances of me being mistooked for a hancient Black-and-Tan Cavalier are not good.

Gran-the-Dog-Hexpert says Mum is not to discourage the lining up and pretending to be Gerry-Hatrick: unless there is a mirror-call, most of the younger Cavaliers will probababbly end up with problems with their arts heventually, so they might as well learn the routine now ...

****FORTS ON TREATS****

- Treats is yummy fings gived to dogs, wot is betterer than most of the fings that they normally get gived to eat
- As far as I are concerned, the honly treats worth having are the ones I get to eat in my bed, on my actual very own with noboddedy looking at me or saying anyfing
- Any other treat is not a treat, it is a trick and I aren't falling for it
- Holding a treat in front of my nose and then actual hexpecting me to move towards it won't work. Hever ...
  ... not heven if it's written in a book. This Lurcher is not for luring
- Very hoccasionally, I will accept a treat outside the house, and you should be actual grateful and pleased I has dunned this fing. This is me being hobliging and rewarding you for being very quite nice and kind. Please do not hexpect me to do anyfing else other than eat it
- It doesn't matter how yummy or how hexpensive it is, if I doesn't want a treat, I aren't eating it
- Heven if the person giving it to me did come a long actual way to see me and might be hoffended
- Wrappering meddy-sins in a treat is a con. Given that some dogs can sniff out boms and hoomans trapped under buildings, I aren't sure who's conning who, hexactly ...
- Kite says she is a Labrador and everyfing I has writted is to be very hignored; she will do anyfing, for anyone, for half a bit of mouldy carrot

SEPTEMBER

September 1

The fuge ginger boyman has got a job! Noboddedy is very quite surprised by that actual fing. He is clever and funny and, just like Mum said, all the fings he did while he was at Universally gived him much more hinteresting stuff to talk about in his hinterview than the other peoples. After-very-all, how many peoples do you know who can do first aid, chop down trees, HAND play the French Horn, as well as know about science and make sure history students aren't a shambles and pay their bills? Dad reckons the fuge ginger boyman is a Jackervalltrades, wot are quite very useful in the real world. Just like him. And he's very proud.

Mum's wants to be proud but before she does gettering to the proud bit she needs to bang her head against a brick wall, she says. But she hasn't moved off her chair so I fink its another one of them metty-fours again, and the reason for the head-bangering words is because the fuge ginger boyman's jeans haven't just created another Jackeralltrade; they've made Another. Blinking. Teacher. He's following a long line of his hancestors into the classroom, starting with Granny Mary and then Gran-the-Dog-Hexpert, followed by Mum and then him.

I dunno about you but I are starting to feel quite actual wobbly and anxious for the fuge ginger boyman's class of childrens. Can you actual himagine wot it would be like to have That. Lot. all in a classroom at the same time? For a start, I dunno how any of the childrens would be able to find a-very-nuff space to speak any words into. And God. Very. Help any of them childrens if they want to do disagreeing with anyfing. Although, finking about it, the Grans and Mum would probababbly be too busy harguing with each other to notice, and noboddedy would learn anyfing apart from how to make a hawful lot of noise.

So I does hope the fuge ginger boyman does decide to just be himself in his noo job at his noo school.

September 2

Good noos! No jeans are allowed in the classroom. So the fuge ginger boyman won't have to do worrying about his long famberly line of teachers, and he can leave all his jeans behind. He will also have to wear a suit and a tie, and can Mum lend him some monies until he gets his first month's pay? The fuge ginger boyman is taking fings all very quite seriously happarently, and he has had an Air Cut. Haccording to the Kat Lady, he does look quite actual different and And Some, and also very growed up. None of this will make any difference to me hunless he has suddenly found a mirror-call cure for his smelly feets. So long as his feets do still smell of six-day-old fish and cheese soup, I will always be able to recognise him, wotever he does with his Air.

104

September 3

I are certain I did just hear Mum say that she is a Dahlia and Kite's Mum is a sacker fishal lettuce. I aren't sure what sacker fishal is but she is quite actual not a lettuce. Lettuces are green fings that some hoomans put in sandwiches. Not Dad, hobviously: he is not a rabbit, he says, and he's not eating lettuce unless he's in very plite comp-knee and he can't actual havoid it.

There was no wine hinvolved in this talking but there was a lot and a lot of flying, buzzing hinsects, and they did all seem to want to eat Kite's Mum. And whilst they were trying to eat Kite's Mum they did not want to eat anyboddedy else who was standing next to her so that made Mum the Dahlia ...

Dad says Mum isn't a Dahlia and Kite's Mum isn't a Lettuce and it's a Nanalogy which sounds like himportant science work that the fuge ginger boyman should be doing but isn't. It's just another very actual hexcuse for Mum to talk endless blinking rubbish about gardening. Again. And if Mum and Kite's Mum stopped wandering around the garden when it's nearly dark, noboddedy would have to be a lettuce or a Dahlia. And could they shut the front door before the house is full of somefing-beginning-with-B-midges ...

September 5

The fuge ginger boyman has dunned his first day at work, and he says it was actual quite knackering, and there is no time for any lunch – or even a wee – but he finks that will change once he doesn't get as lost as he did today. The honly other fing he is struggling with is his name: all the childrens do have to call him Sir or Mr fuge ginger boyman, and it sounds weird and he keeps finking they are talking to somebodeddy else,so is lookering behind himself for the grown up they is speakering to.

Mum's wondering how she did get so old so very fast ... and wondering how long it will take the fuge ginger boyman to realise that halmost every single hadult in the hentire world spends most of their actual lives looking behind them for the grown up ...

September 6

We have had the saddest noos here. Granny Wendy has died. And heven though we has been actual hexpecting this noos for some time, everyboddedy is still feeling hupset and shocked. There are some fings you just can't get ready for, however much you try.

Everyboddedy, apart from me, will be going to Corny Wall in a few days to say goodbye at a special service called a funeral, and the fuge ginger boyman has been gived the day off from work so he can go as well. Fortunately, he does already know where his suit is and, as he's honly actual owned it for about a week, that it still fits him. Until about ten minutes ago Dad did not know neither of these fings about his suit. Now he does know where his suit is and he also knows that the trousers won't fit him unless he does leave all the buttons undone or breathes in a lot and a lot. Fortunately, there is a waistcoat bit of the suit that goes over the top of the trousers, so Dad finks he will just a-very-bout get away with leaving the button undone ... and hoping that his trousers don't fall down.

For THE QUITE ˅very ˂actual LOVE of Worzel Wooface

September 7

Mum says Dad Will. Not. be wearing nearly-falling-down-trousers to say goodbye to Granny Wendy, and she's moved the button on Dad's trousers as best she can. He'll still need to wear the waistcoat to cover up Mum's very actual bad stitchering, which made Mum smile and sad at the same time. Last Christmas, Mum made a cushion with a fuge flower on it for Granny Wendy, who was very frilled with her present and quite actual himpressed. The cushion tooked Mum for hever to make, and although it looked neat a-very-nuff on the outside, it was a right blinking disaster hunderneath, just like Dad's outfit for saying goodbye, and he's not to do eating too much or bend down too suddenly. Finking about sewing and Wendy seems fitting today, Mum reckons, which is more than can be said for Dad's trousers ...

September 8

I will be staying with Gran-the-Dog-Hexpert again when my famberly goes to Corny Wall. They won't be gone very actual long, and a six-hour car journey without a long gap before another six-hour trip home would be too-much-for-Worzel. And although goodbye services can sometimes be special places for dogs, Granny Wendy was a very pop-oo-la person with a lot of hobbies and hinterests, so there will fundreds of peoples there and standing room honly. Peoples will not want to be worrying where there feets are and if they is treading on a luffly boykin whilst they do some singing. So, all-in-very-all, I are quite glad I will be staying in Suffolk, hespecially after the last trip and the actual rubbish-and-not-actual-very-friendly 'dog-friendly' hotel. Or maybe too friendly in the wrong direction, with all the fleas that decided to be my Noo Best Friends and come home with me.

While I are finking about that fing ... here is the Worzel Guide to Dog-Friendly Cafe and Hotel-doing:

****FORTS ON DOG-FRIENDLY****

- It is not a-very-nuff to stick a sign on your door to say that your Hotel is Dog-Friendly – you does have to actual do somefing to make it that way
- Dogs do come in all different shapes and sizes. I do realise that I are ignormous but I would like to be able to lie down hunder the table just as-very-much as a Jack Russell Terrier
- Grapes is actual poisonous to dogs. Mum would rather not have to grub around hunder the table scrabbling to find the ones wot have fallen off her helegantly deccy-rated cheese and biscuits ... ... and then discover that there is a mouse trap with poison in it down there as well
- On the motorway when Mum and me is travelling alone, she does sometimes need a wee. Hurgently. Happarently, I are either coming in with her or she is peeing on the grass. In broad daylight ... it's your choice
- It would be quite very helpful if you could provide a bowl of water for me to have a drink from. Gettering my face in a teacup is a quite very actual challenge, and possibibbly not that hygienic
- The route to all-very-hell and chaos is the dog-wot-owns-the-pub wandering freely around lots of other doggies wot are all on leads
- Spray. The. Carpets. Hespecially in the summer
- Please can we have a Nin and a Nout if possibibble? Or a bitta glass that hoomans can peer through to check what's coming and going. That way there will never be a hunexpected and

too close hencounter with another doggy at the door

It is perfickly possibibble to have too much of a good fing. 47 dogs in one cafe is not Dog-Friendly - it's a disaster waiting to actual happen

September 10

Everyboddedy is back from Corny Wall and has mixed forts about the trip. It was booful at the place where my famberly said goodbye to Granny Wendy, and so many people came that there was nowhere for the fuge ginger boyman to do sittering down. There was an ignormous window that looked over fabumazing rolling hills, and Granny Wendy choosed the place herself because she did fink everyboddedy would like the view. Mum says she was quite actual glad I was not with her because I would have fort it my Himportant Dooty to run around all over Granny Wendy's fabumazing rolling hills, and she would have spent the rest of the afternoon trying to get my tracker to work in a place with halmost no hinternet.

Afterwards, there was a party at the bowling club where Granny Wendy spended a lot and a lot of time. She used to pretend to be an Armless Little Lady at bowls but that was a very lie. She did bowl-rolling for the county, and could beat just about anyboddedy who was stoopid a-very-nuff to fink they might like to play a game with her.

Dad says it was actual quite strange to be in Corny Wall without Wendy there to quietly keep Granddad horganised, and make sure he was eatering proper food and not just fings that can be heated up in the machine-that-goes-ding. Auntie Sue is going to be very In. Charge. of that now, and she says she is not lookering forward to it. Granny Wendy was a Force-Tubby-Reckoned-With. I can see it gettering quite actual noisy in Corny Wall over the next few weeks, and I are quite actual glad I do live in Suffolk which is about as far from the you're-going-to-turn-into-a-chicken-korma words as you can get in Ingerland.

Dad says I are not the honly one ...

September 12

We went to see our friends, John and Liz, at The Ship Inn on Thursday. We did drive all the way to nearly Hipswich cos they is special and himportant friends who we don't see a-very-nuff. I fink they does fink the same fing about us cos they did drive all the way in the oppy-sit direction to see us ... to the The White Hart in Blythburgh ...

Dad says Mum is never, hever allowed to say men is rubbish at communercating again. Hever. Mum and Liz were very In. Charge. of the arrangements, and Dad and John were honly in charge of driving and wearing-a-clean-shirt and doing-as-they-were-told. Mum says there is nuffink a-very-tall wrong with her communercating. Just her memberry. And Liz's is a bit flaky as well. It was a perfick storm of forgettering. So when she and Liz very harranged to meet-at-the-pub-we-went-to-last-time, Mum forgot about the trip to the White Hart. And Liz forgot the name of the Ship pub, so it was a complete and hutter, erm, cock-up for the hoomans. But it was fabumazing for me!

I did get to have a super cuddle with James-the-Landlord at the Ship, who does know hexactly where I do like to be scritched. Then Dad and me went for a proper long walk down by the river whilst Mum and Liz did huntangling the mess and muddle of peoples-being-in-different-pubs-miles-from-each-other. And Dad got to say all of his stoopid-somefing-beginning-with-B-woman words out of the way of Mum's hearholes.

September 13

All day today we have had. Men. In. The. Garden. Louis and his men have come to do choppering back the ivy around our windows because our house is starting to look like Sleeping Booty's Castle. Happarently. It has been my himportant work to do lettering them into the kitchen to make their own cuppateas because Mum says she has got work to do, and that way they will be able to have what they need when they want it, rather than relying on Mum to remember they might be dying of first.

We have a rooteen for lettering people into our kitchen, and everyboddedy has to do as they are told so it works perfickly, hespecially the peoples that want to use the kettle so that they can come-and-go-as-they-please. All they has to do is put up with feeling a bit very redickerless for five minutes. Dad says it is hembarrassing and God. Very. Help. Us if we do have to have someone himportant at our house, but Mum reckons if fuge great police people can tolly-rate it so can everyboddedy else ...

****HOW TO BE A VISITOR IN OUR HOUSE****

- Come to the house and get hambushed by Mum in the front garden, and get tolded to come to the back door and don't-forget-to-shut-the-front-gate
- Come in through the back gate and don't forget to shut that one neither
- Get 17 napples land on your head cos the back gate wobbles the napple tree
- Say Ow!
- Run into the house being chased by a billion wasps
- Listen to Mum talk at you like she is presenting a children's telly programme about how-wonderful-to-see-you and we-won't-have-to-do-this-for-very-long. In a sing-songy voice that makes Dad hide upstairs in the bedroom. And cringe in hembarrassment
- Very allow me to do hobserving all this through the cat flap in the kitchen door and see that you is a super, fabumazing visitor who is probababbly Mum's bestest friend ever, hever, and quite very actual someboddedy wot I would love to say hello-I-are-a-luffly-boykin to
- Stand habsolutely still, with your hands in your pockets and look out of the window whilst Mum opens the hall door to let me into the kitchen
- Hignore me very completely whilst I do my bestest wiggling up to you and sniffing you, huntil I do nudging you for a fuss
- Get added to my list of Peoples-I-Like and make yourself as many cuppateas as you do want

Mum says she has gotted habsolutely No. Hidea. if all this is necessary, and the chances are I would do running away if someboddedy came into the house and we didn't do this gettering-to-know-you-rooteen. But this works and keeps everyboddedy safe and happy. So if you want to come into our house then you have to do puttering up with it.

Dad reckons Mum is redickerless, and I are perfickly capababble of coping with noo peoples coming into the house without her acting like a Total. Plonker. Either that or it's just a cunning plan to actual discourage visitors.

September 14

The Men. In. The. Garden have dunned an exerlent job with the ivy, and our house does look smart and not-like-it's-been-abandoned or left to get swamped like in a fairy tale.

It was about time to be very actual honest. In fact, it was probababbly about time a year ago. Last year Dad did have a go at it but he isn't keen on ladders or gardening, so he just got the stuff away from A-Bit-Deaf-Saint-Fred's hairy-hall and the chimberly. Saint Fred, who lives next door to us, gets hupset when his telly don't work, so Dad fort it was the least he could actual do, seeing as he has-to-put-up-with-us-lot. Dad fort the job was dunned but then he remembered why he hates gardening so much. It's because nuffink *stays* dunned. It's not like boats. When you do a job on the boat, it stays dunned, he says.

But plants do growing back and need choppering Every. Blinking. Year. and Dad says his knees are too sore for ladder-climbing. Though not for sailing. Sailing uses a completely very actual different set of knees, I do fink. Dad has one set of knees for sailing and another set for ladder-climbing and dancing. And it's the dancing ladder-climbing ones wot are broked. Dad is very pleased with Louis' ivy choppering because he got it done in No. Time. and he-had-the-right-tools and he's a Fit. Young. Man.

Mum says Louis isn't a Man, he's an eight-year-old ickle baby boy – at least he was the last time she saw him. It was all a bit of a shock when he pitched up on the doorstep to do the ivy choppering. The last time Mum saw Louis he was in the fuge ginger boyman's class at primary school ... and he's grown. A-very-lot. Now he does runnering his own gardening business, and has two other actual men who do working for him, and somehow he has got old a-very-nuff to be allowed to use noisy, screechy hequipment. And climb up a ladder. Mainly because he isn't eight years old anymore, Dad reckons, he's a Grown. Man. and probababbly a very quite hembarrassed one at that, seeing as Mum spended all afternoon remindering him how she taughted him fractions ...

September 16

Mabel has got a fuge ouchie on her ear. It's like someboddedy has stuck a straw between the bits of skin and then blowed into it so it looks like a balloon. Dad is in the The Nile as actual very usual, and muttering words about it'll-be-okay and she-doesn't-need-to-go-to-the-Vet. After Mum had called Dad all sorts of rude and perfetic words about him being Vet-fobick, she hannounced that Mabel didn't need a vet but she did need some First Aid. At which point, Dad did look Orry-fried because not honly is he actual Vet-fobick, he is also completely very against any cat-fiddling that might cause them the slightest actual bit of uncomfy-ness. Heven if they does need it to feel very better in the long term.

When there is cat-fiddling to be done, Mum says it is my himportant work to do bogging off. Heven if it wasn't my himportant work, I would do bogging off. Big-Time-Badly. With knobs on as a volley-on-teer. Dad also tried to volley-on-teer to bog off but he can't move as fast as I can, so did get shut in The Hoffice with Mum and Mabel, at which point he did discover that it was his hessential job to Hold. On. To. Mabel. whether or not he liked it. And keep-her-still and I'm-not-hurting-her-you-useless-man-stop-saying-that.

After Mum had removed the cat claw sheaf wot was stucked in Mabel's ear and gived Dad her bestest told-you-so look, Mabel's ear did return to halmost its normal size, and she was actual allowed to hescape to her shed to do licking off all the cream Mum put on her ear to stop it getting a hinfection. Dad would also like to go and hide in the shed but first of-very-all, he's gettering a Talk About Responsibibble Pet Ownership. And some cider ...

September 17
Dear Wailing Car Halarm
Please stop going off at half-past six in the morning. You aren't being robber-dobbed. You is just being actual noisy and hencouraging me to bark, and making Dad to miss out on an hour of sleep and Mum be really, wheely grumpy.
From your luffly boykin
Worzel Wooface

September 18
Dear Wailing Car Halarm
Fank oo for not going off at half-past six this morning. Going off at quarter to seven and then every ten minutes after that was not that much betterer, though.
From your luffly boykin
Worzel Wooface

September 19
Dear Person Wot Does Own The Car With The Noisy Wailing Halarm
Mum and Dad do have a fuge shed with tools in it. They is gettering less and less hafraid about using them.
From your luffly boykin
Worzel Wooface

September 21
Dear Very Clever But Noisy Starling
Fank oo for actual revealing yourself as the Phantom Car Wailer before any hammers gotted used or neighbours shouted at. Standing on our roof and pretending to be a car halarm is Not. Funny. Any. More. Although we is all quite actual himpressed that you have learned how to do this fing.
Please could you either go back to being a Starling, or find somewhere else to pretend to be a car halarm. Or just do your showing off a little bit later in the day.
From your luffly boykin
Worzel Wooface
Pee-Ess: Gandhi says ... well, nuffink. But I fink you should know he actual hexists.

September 22

I made some noo friends this weekend. We have been to visit Roxy and Flint, who are Whippets. In very general, I do sometimes find myself on the wrong end of careful-Worzel-you-great-oaf around Whippets, so I was not sure how our visit would go.

Whippets are confuddling. To hoomans they look like delly-cate and precious Fings That Break easily. To dogs, they just look like dogs, but because hoomans have delly-cate breaking forts around them, they can sometimes get away with stuff that fuge, ignormous Lurchers cannot. Or it might be that because they is small, noboddedy notices until it is too very late.

I was on my bestest behaviour, and did not squash Roxy or Flint, but I did have a fabumazing time roaring up and down their long, thin garden. It was my himportant work to do this and hentertain the hoomans, so that Flint could use the distraction to crawl under the fence and into the field next door, which was full-of-rabbits and he'll-be-gone-for-hours-now. Happarently, Flint is a fusey-tastic rabbit hunter but not too actual talented. Roxy has mainly retired from rabbit-chasing but she used to be exerlent at it. Nowadays, she is more hinterested in lying in the sun.

I did not get the chance to find out how good I could be at rabbit-chasing. Mainly as I couldn't crawl under the fence without everyboddedy noticing.

How-very-ever, in the afternoon, we went for a long walk on The Downs. Calling them Downs is quite actual odd I do fink. I are wondering if it is a trick to get peoples like Dad, who have very actual knackered knees, out for a walk. Downs are not Down. They is Up. In very fact, they is the uppiest bit of countryside in the hole of Wiltshire I do fink, and it did feel like we were walking on the top of the very world.

I are pleased to say that I did manage to do a small amount of bogging-off, but not Big-Time-Badly, and I wored my tracker. The Gee-Pee-Ess signal on the Uppy-Downs wasn't very actual strong, which Dad did wonder was somefing to do with the soldiers who do practicing near there with somefing called live-ham-you-nishon. All the talk of live-ham-you-nishon did nearly give Mum an art attack. There is clearly marked-off areas, happarently, but Mum did decide that she would still like to have an art attack, as she wasn't that convinced I would be able to read the signs or pay much attention to the flags ...

In the end, Dad tolded Mum to stop being redickerless, but I do fink her worrying did actual hencourage me to stay close. Either that, or the fact that I was on the top of the world in a strange place where I didn't know any of the sniffs meant I had to use my nose a lot and a lot. And it is halmost himpossibibble to use my nose to do learning a noo place and at the same time, let my legs do bogging-off.

When we got back from our walk up on the Downs, I was quite actual hot, and if I hadded been at home, I would have headed straight for my paddling pool. Flint and Roxy's house didn't have a paddling pool, and I has some fuge Complaints to the Management.

For THE QUITE ̌very actual LOVE of Worzel Wooface

Roxy and Flint have Got. A. Stream. In their garden. Wot is completely and hutterly Wasted. On. Them. cos they is Whippets. and everyboddedy knows most Whippets will dip their toes in water but aren't at all actual hinterested in splashing and plunging and gettering completely very over-hexcited about A. Stream. In their garden.

So I do fink that Dad and Mum should do offering-to-take-it-off-their-hands and bring the stream back to Suffolk. With Us. Where I will do luffing it every quite very actual day and help to make it feel special and himportant, and Not. Hignored by Whippets.

September 24

Doris, Kite's older cat, is poorly-sick. Today, she has been listless, wot is so hunlike Doris it is himpossibiblle to very actual hignore. Doris doesn't do calm or relaxed or any of those fings that could get muddled up with listless, so it was easy for her Mum to see that somefing was wrong. She was also very quiet to go along with the listless, and for a small moment I did fink I might like the noo-not-noisy-and-not-bonkers-crazy version of Doris, but happarently, I aren't actual allowed to fink those kind of forts. Heven if she is the honly cat wot has ever decided to launch itself at me and try to start an hunfair fight. And not hunfair because I are at least ten times biggerer than her, but hunfair as in she-started-it, for habsolutely No. Reason. A-very-tall. Apart from the fact that I are a dog and she is redickerless and bonkers and Not. Safe. For. Worzels. In. The. Garden. Nowadays, we aren't actual allowed to hencounter each actual other; it's just not worth losing a hiball over.

Haccording to Kite's Mum, Doris is perfickly luffly hinside their house but outside she is a different cat. So different that she really doesn't act like a cat a-very-tall, but more like a furious and hungry and suddernly woked up Tie-Ran-He-Saw-Us Wrecks.

So Doris has gonned to see Boris-the-Vet so he can decide wot is wrong with her and how it can be fixed. I are wondering if I should give him some pointers.

September 25

Doris is still actual listless and has had to stay overnight with Boris-the-Vet. Haccording to him, she has got quite very actual hawful die-a-rear, and for a Vet to say that fing it must be really, wheely bad. Kite's Mum says she is missing Doris a lot and a lot, but when she hearded about the die-a-rear she decided she wasn't missing her a-very-nuff to fink about offering to do home-nursing.

BOOTS

It doesn't matter how quietly
You slide your boots onto your feet
I will always hear you
Even when I'm fast asleep

I'll hear you from the bedroom
I'll hear you over the noise

Of the TV and radio blaring
Or the singing band of boys

Heven if you fink it's boring
And not somefing a dog wants to do
If your boots are going on your feet
Then I'm coming along, too

continued page 121

Doing careful saying hello to Nellie the kitten.

Hattie-Hazel Princess Poo Pants has growed up to be very quite booful ...

... and actual clever, too!

Kite helpering Harry's Dad to do photo-ing.

Playing tuggy
with kite is
easy-peasy ...

... but playing
tuggy for the
first time with a
hooman did take
a lot and a lot
of finking about.

Harry is a quite very actual
mooze for his photo-ing Dad.

Sometimes I are helpful
in the garden ...

... and sometimes I are not quite very actual helpful at all!

The fuge ginger boyman has grab-you-lated!

The very clever seagull with the fuge ginger boyman and the kat Lady.

THE WARRIOR WALK

Darwin's baby.

Roxy and
Flint have a
fabumazing
stream in their
garden ... I do
fink I are missing
out big-time-
badly ...

I had to have a
bath ...

... so I could
be an exerlent
prop for a telly
programme.

47

Reserved
Worzel

Some pubs are
betterer than others
at being Dog-friendly.

I sleep on the big bed.
Or I do sulk.

Me and Kite and Maisie,

There are only so many cats
that can do sitting on Dad

Which wasn't a fabumazing hidea to be quite actual honest, as all Mum was doing was going to the garden and the shed to throw stuff about and smash fings into tiny pieces so they would fit into the bin. And then chopper up some bits of twigs using Kite's Dad's bit of machinery wot sounds like it is eatering stuff really, wheely noisily, and I-did-tell-you-Worzel and do-you-want-to-go-back-in-now?

Mum needs to get two pairs of Wellies, I do fink, like Dad has got two sets of knees. One set for walking in fields with Worzel Wooface and another for doing nasty, scary, horribibble jobs in the garden. Then I will be able to tell the actual difference and not find myself in situ-nations I would rather actual not. She says she used to have another pair but they must have hoffended me and got very hassociated with noisy machines and shed-hurling.

Cos I eated them ...

September 27

Doris is still actual hunwell, and there is some finking that she might have binned poisoned, which is Orrific and frightening, and not only for Doris. Everyboddedy in Kite's house has had a proper good look round and they can't find anyfing that she might have haccidentally poisoned herself with. We are all hoping that it isn't somefing more Sin-A-Stir. If Doris has binned poisoned it could mean that all of the other cats near where we live might be in danger as well.

Everyboddedy else in the area is being sensibibble and waiting for the test results from Boris-the-Vet, but Mum has decided that there must be a Phantom Cat Poisoner in the village. All the cats are under House Arrest and being actual hinspected every hour to make sure they isn't being listless.

I are wondering how we will be able to tell the difference between a poorly-sick Frank and a normal, nidle Frank. And also finking that if Frank is going to be ill or poisoned, he needs to fink of somefing else other than being listless to show that he is sick. He never moves off the kitchen table, and the honly fing about him that seems to have any get-up-and-go is his tail. The rest of him just lies-down-and-sleeps.

None of the other cats are lookering listless. In-very-fact, they is looking cross, and not happreciating Mum's hinspections and concerns. A-very-tall.

September 28

I knew I should have gived Boris-the-Vet a list of all the fings that are weird and just actual wrong about Doris. Not with Doris, hobviously, because that is Boris's job, but someboddedy shoulda dunned pointering out to him that Doris walks in a way no cat that wants to be considered normal should walk ,and more like a Tie-Ran-He-Saw-Us Wrecks than you woulda fort possibibble. Doris' front legs are very actual shorter than her back legs. Kite's Mum reckons Doris' Mum and Dad were already related to each other before they did become her Mum and Dad, so Doris doesn't have a-very-nuff cat jeans, which is why she is like a Tie-Ran-He-Saw-Us Wrecks. I aren't completely actual sure about this feery, but I do know that Doris walks like there is bits of bone missing from her front legs ... and like Mum does when she has binned drinkering wine.

For THE QUITE ˅very actual LOVE of Worzel Wooface

But now that has binned cleared up, Boris has dunned some other tests and it turns out that Doris has got a nasty hinfection inside of her, and there isn't a Phantom Cat Poisoner on the loose. Doris will get betterer and, most himportantly, as far as I are concerned, Mum has opened the cat flap so that our cats can go outside again. None of the cats were wot you'd call patient or frilly-sofical about being kept hindoors, and their forts and feelings about this were beginning to be very actual hexpressed in a way that noboddedy would ever describe as listless ...

September 30

Fings are very quite stinky and smelly at our house at the moment. Mum says it isn't Dad or my bottom burps but it is Somefing In the Hair. It is giving Mum a Ned Ache wot is very not good news cos she does not want to go outside to play with me. Or go for a walk. And when I wented round to Kite's house this evening, she could only cope with about twenty minutes before her hiballs started to water.

There is honly One Fing For It as far as I are concerned. If it is Somefing in the Hair, then the Hair has got to go. And she is going to have to have it all shaved off like Dad does every year. I does realise she might look a bit actual strange with No Air, but it is all a-very-question of Pri-Orry Trees. Walks and playtimes is actual very quite himportant and hessential. Air. Is. Not.

Dad says, fortunately, I has got it all wrong and there is no need for him to start brandishing about his Air Clippers. Hespecially as last time he got the settings all wrong and cutted his air so blinking short it took weeks for him to not look like, well, redickerless or a convict or any of the other fings Mum shrieked at him. She refused to be seen in public with him for ages until it growed a bit. The smell, Dad says, is cos someboddedy has binned doing Muck Spreadering, wot means a farmer has been flicking poo all over his fields to help his plants grow.

So now I are even more actual very hoffended: Mum is refusering to take me for a walk until the smell goes, wot means I aren't going to be able to hinvestigate Any Of The Poo. Wot is my himportant work and very actual dooty as a luffly boykin. So now I are sulking. And Mum is sneezing. And Dad just wishes everyboddedy would remember to shut the blinking door before the hentire house smells of Chicken's Hit.

OCTOBER

october 1

Every morning Mum gets up
And leaves a warm spot in the bed
From where her bum has burped all night
(I don't fink that's wot I shoulda said)

Erm ...

Every morning Mum springs out
Of bed to greet the luffly day
And I budge up to feel the warmth
Of where her sleeping boddedy lay

    Currently, none of them fings are actual true. Well, the budging up bit
is, but none of the rest of it is. There is no springing and no sleeping, and if the
day is luffly, Mum isn't aware of it. All she's noticing is that the night is over and
she's not had any sleep ...

october 2

Dear Mum
If you did want a luffly boykin who would curl up into an ickle ball and not do hogging all the actual
bed, you shoulda got a Yorkshire Terrorist.
From your fuge, ignormous designed-to-do-stretching-out Lurcher and not an ickle doggy.
But still a luffly boykin
Worzel Wooface
Pee-Ess: I does not know why you do need to feel your feet. You don't do really, wheely a-very-nuff
runnering about to need to feel them any-blinking-more than you already do ...
Pee-Pee-Ess: If you do want a bit more room, the hall floor is very actual comfy, as you keep saying ...

october 3

Yesterday was Hinternational Coffee Day. Mum says it didn't help. A-very-tall.
And she is really, wheely tired and somefing will Have. To. Change.
    There do seem to be a very lot of 'days' for stuff. There's heven a
National Cabbage Day in February. I aren't kidding. And in April there is a
Hinternational Kissing Day. I are starting to fink peoples are just lookering for
hexcuses to wear hats and drink too much wine, to be quite actual honest.
    Mum says she has had a-very-nuff, and today is going to be Worzel
Sleeps Somewhere Else Night. Hand Day, whether I like it or actual not, and
there is nuffink I can do to distract her from this Orrendous Fort. She says she
has had a-very-nuff of wakering up every morning with numbed toes. Tonight,
she is Shuttering. The. Door. so she can get at least one decent night's sleep:
National I Love My Feet Day was in August so it's already gonned and Mum

says she is Not. Waiting. for another ten months to Sert. Her. Rights. to not have numb feet cos Worzel has binned lying on them all night. Again.

Hunfortunately, I've missed National Pet Day as that isn't huntil next April, so I can't heven suggest that as a reason to do puttering this off, and I are running out of Hideas. So tomorrow, it is going to be Worzel's Sulking Day and I do suspect I will be celly-brating National Sofa Day early. The actual National Sofa Day isn't until 6 November so I could be stucked on there for a hole month ...

october 4

Dear Dad

Please tell actual Mum that the hall floor is Not. Comfy. I founded this fing out last year when Harry did such fuge ignormous bed-hogging that it meant I had to give up sleepering in the big bed as-very-well. Or else he did actual howling. The hall floor has not himproved a-very-tall and I doesn't like it. Heven if it does have a memberry foam mattress **HAND** a duvet ... it's Not The Bed. The Bed is where I want to be. So, as long as I are not On. The. Bed. I is Not Comfy.

From your luffly boykin

Worzel Wooface

Pee-Ess: I fink if you doesn't stop snoring soon, you'll be finding out just how Not. Comfy. the hall floor is, too ...

october 5

I did sneaky bin-raiding. Bin-raiding is somefing I aren't knowed for, so I did gettering away with it. Nearly. Hunfortunately, the old coffee wasn't as actual good at being sneaky as Worzel Wooface, and did giving me away by scattering itself All Over The Floor. And gettering stucked in my beard. So I was Discovered. But not until after I had eated the bitta old bacon I did find.

In very general, I do find that I can get away with quite a very lot of fings other dogs get Words said at them about. Mainly cos I do so rarely try to get away with anyfing. I are too cautious for my own actual good, Dad reckons, so anyfing that is new that I try is A Good Fing, heven when it is a Bad Fing. Like raiding the bin. Lots of the foster doggies wot have stayed here have tried to actual hencourage me to join in with it; it is their favourite hobby and one they fink I really, wheely should be better at. And quicker. But mainly join in and stop standing by the door looking worried and wobbly and givering them away ...

Other famberlies might fink mine is a bit actual weird and strange for not getting cross with me for raiding a bin, and say that I are hindulged or spoiled, but that is because they do not know how brave I do have to be to try somefing noo. Heven if it is somefing that isn't to be actual hencouraged again. Just trying somefing noo is exerlent and that's as far as it goes.

Currently, as far as Mum is going is the sofa, even though most of the old coffee stuff is all over the floor and the rest of it is stucked in my beard. She says she's got a Ned Ache from not a-very-nuff sleep, and Dad can deal with it when he gets home.

october 6

Mum's come up with a feery. It is the Feery of Dad's Snoring and she says that now she has got to test her Hi-Pothy-Sneeze. Or somefing nearly like that; I aren't sure ... But it was somefing the fuge ginger boyman did do at Universally and it's all about when you come up with a Hidea and then you has to actual prove it. Or not prove it. Not proving a Hi-Pothy-Sneeze is also very okahy too because it means the science people don't have to worry about finking about that hidea any very actual more.

Mum says, That's. Fine. for sigh-and-tists but not for Mums. And my Mum in very actual part-tickle-ar. She really, wheely needs for her Hi-Pothy-Sneeze to be right, because otherwise she's going to Chop. Dad's. Blinking. Head. Off. I do fink that this might be actual going a bit far ...

Mum's Hi-Pothy-Sneeze is that Dad does snoring when he is tired. Now, I aren't a sigh-and-tist, I are a dog, but heven I can see that it is pretty flipping hobvious that Dad snores when he is tired cos he honly does it when he is asleep ... and sleeping is what anyfing with a art beat does when it needs a rest. So I aren't completely sure Mum's feery is going to hextend the front-ear of science very far. And neither does Dad. Dad reckons Mum's gonned mad. She needs to have a nice sleep in a dark room, he says ...

It's been a quite actual long time since Dad did get That. Look, and I aren't even sure whether he did see it; he was already on his way to the shed to dig out his sleepering bag from all the sailing stuff. I fink he knew before he'd finished sayering his words that he'd be sleeping on the sofa tonight ...

october 8

All this week, lots of the talk during my playtimes with Kite has binned about the white bobbly fings wot have suddenly popped up all over Kite's lawn. At first, the talking was about whether Worzel Wooface would fink they was toys or, heven worser, hedible, and then there was a lot and a lot of worrying about wot was to be dunned about it. I can tell you for-very-nuffink, I do not fink they is toys or actual food. I does know they isn't food cos first of-very-all, Maisie, Kite's sister, wasn't eating them, and she eats everyfing, and second of-very-all, they isn't chicken wings. And I do not fink they was toys cos Kite wasn't trying to play with them.

Once it had been actual hestablished that the-dogs-wouldn't-be-so-hidiotic, all of the hoomans did have to work very actual hard at not being stoopid as well. That was much, much very harderer, and you would not believe the bad hinformation and Folk Lore five peoples can come up with when they has had a glass of sherry, none of them know wot they is actual talking about, and there is a danger that Free Food might get wasted. And there's nuffink on the telly to distract them.

I are actual very pleased to say that my Mum did leave all the white bobbly fings where they was. She says she knows nuffink about mushrooms. They is probababbly field mushrooms but she's got far too much to blinking do at the actual moment to have bellyache or a cycle-delic trip to the moon.

When we wented round to Kite's tonight they had all gone. You will be quite actual relieved to know that all of Kite's hoomans are still alive and well, and haven't gone on a Magical Mystery Tour. But I are a bit actual worried about Kite's Mum's lady-wot-Voovers, cos she's taken all the white bobbly fings, and happarently she did come up with heven more Folk Lore wot didn't make any sense at all. Mum finks Kite's Mum shoulda stopped her: apart from the bellyaches and the hintergalactic tours, she might do dying.

And everyboddedy knows it's actual quite himpossibibble to get a cleaner nowadays ...

october 9

Mum's been doing more talking about her Hi-Pothy-Sneeze tonight, and this time Dad has had to do listening and not just fast forward to the bit where he ends up sleeping on the sofa. Mum has decided that her Hi-Pothy-Sneeze isn't just a-very-bout Mum getting a Good. Night's. Sleep. For. Once. but is also about Dad's Elf; like eating vegetababbles and wearing sun cream. Dad is not hinterested in eating vegetababbles. A-very-tall. So he did decide to do listening to Mum's feery and Hi-Pothy-Sneeze. I fink he's hoping he can swap the vegetababble nagging with the Hi-Pothy-Sneeze and Mum will stop hiding carrots in the Spag Bol ...

Anyway, Mum reckons that when Dad is really, wheely tired because he has binned having lots of hard days at work, or it has been a bit cold or windy down at the harbour, or he has stayed up actual too late playing on Uff the Confuser, then his snoring is actual very much worserer.

So, to test Mum's feery, Dad is going to have to go to bed earlier. A hole hour earlier to see if it makes any difference. And then Mum will come to bed at the normal time and see if Dad's snoring is any quieter. Which was all very well and actual good, in feery, but, by the time Mum came up to bed, she did find that the bit of the bed she usually does lie down on was quite actual hoccupied by a luffy boykin, who does fink that this hole science fing is quite very fabumazing.

For the first time in a week, I has managed to get to sleep on the bed. And not on the hall floor. Mum can sleep on the sofa and I will do sleeping next to Dad and do reporting back to Mum in the morning. Or not. Cos I do have to say that Dad's snoring does not bother me a-very-tall. Mum's Hi-Pothy-Sneeze might not be proved but I fink we has found a Hole Nother Solution ...

october 11

Fings have binned quite very loud in our house this evening. The poxy guv'ment has dunned an hinspection of the Men-Tall Elf services where we do live, and has founded out they is the worst in the hentire country, and that's hardly-a-surprise-is-it? Well, it probababbly was for the Men-Tall Elf services cos they is clueless about everyfing, Dad reckons. The man on the telly said that people in charge of the Men-Tall Elf services were 'disappointed' about the results. Wot started everyboddedy off again being heven louderer. From the hexploding shoutering at the telly, I fink disappointed was the wrong word to choose. And how-about-ashamed-or-saying-sorry? Dad reckons there is Fat. Chance. Of. That. and perhaps everyboddedy should stop yelling now cos it's really, wheely loud.

And Worzel Wooface would like to do coming downstairs.

october 13
****FUGE NOOS!****

Mum and the previously ginger one have binned actual hinvited to do shoutering *On* the telly news. Not just *At*. It. And not honly on our local telly but on the one that gets showed to the Hole. Of. Ingerland.

And wot is heven betterer is that they want to do filming it all down at the harbour so Dad will be able to watch Mum gettering furious and being hembarrassing in front of all of the people he does work with. He's frilled. But not as frilled as Mum who has binned practising some of the words she wants to use by yellering at the telly again. Happarently, the man In. Charge of the Men-Tall Elf services has resigned, and why-didn't-he-do-that-years-ago?

october 15

Today, I did decide it was my himportant work to make sure Mum didn't have fuge actual hystericals on the telly in actual front of the Hole. Of. Ingerland. I did do this fing by being a luffly boykin and very With. Mum down at the harbour whilst they did the filming. I did do a fabumazing job. Dad reckons I do turn Mum's volume down in ways he has never been able to.

The previously ginger one did say a few words about living with a Men-Tall Elf, and then Mum did say most of her telly-shoutering words around the feem of disgustering, disgraceful, knackered, and happalling. Most of the pictures on the telly were of the previously ginger one, which was a good job, I do fink, because she did make a fuge actual effort to look smart and neat. Mum was too busy trying to remember all the ruining-people's-lives words, and finks she probababbly shoulda dunned somefing betterer with her air. Happarently, the film will be on the noos tomorrow night at six o'clock, and millions of people are going to see it.

october 16

This morning, the lady from the telly noos did phone Mum to check that she was still happy to be on the National Noos. Mum managed to say yes about seventy different ways. With knobs on. After Mum had dunned saying yes for about ten minutes where one second woulda dunned, the lady ranged up the previously ginger one on her mobile phone to ask her the same fing. Mum was very actual himpressed, and the previously ginger one was, too, once Mum had hexplained about Dooty of Care and the Bee-Bee-Cee wanting to make sure she was happy in her own head with it all.

Tonight, we watched the noos at the same time as everyboddedy else in the Hole. Of. Ingerland. I has discovered that the quickest way to stop my famberly yelling at the telly is to stick them on it. They did all watch in stunned and very actual silence. Heven though all of their mouths was open, No. Words. Came. Out. I did not find Mum being on the telly as confuddling as my famberly did fink I would. But then, after the last few days of telly yellering, the Stunned. Silence. was more than a bit actual weird, and I was far more hinterested in hobserving my famberly in the sitting room being tents and wobbly than watching the ones on the telly being their normal houtraged selves.

For THE QUITE *very* actual LOVE of worzel wooface

After it was over, Mum did a bit of crying and Dad did a lot and a lot of hugging ... and then all of the phones in our house had hystericals and did make up for all the silence by dinging and bleeping and ringing and pinging for the next hour. Everyboddedy has said Mum and the previously ginger one did do a fabumazing job but the fuge ginger boyman finks Mum might have over-dunned the hexhausted carer look, and was that why she didn't brush her air?

october 17

I has learned a noo trick! It is quite actual rewarding for luffly boykins so I does reccy-mend it for any other sighthound wot is more on the hezzy-tent side of sertive.

My noo trick is Oppy-Mistic Hanging Around. Oppy-Mistic Hanging Around is Not. Begging. Begging hinvolves doing sit and looking hopeful and being a hogpig Labrador. Oppy-Mistic Hanging Around is best dunned when roast chickens is being hacked about by cross Mums. Mums wot is cross cos they asked Dad to come and carve the chicken cos she is rubbish at it, and also trying to make gravy and drain peas and will-you-take-those-blinking-headphones-off-I-don't-care-how-many-people-will-die-it's-tea-time. This is the point when a luffly boykin should do Making A Happearance. Hastily-hacked-about chickens are perfick for luffly boykins cos the skin becomes a casual-tee of bad hacking, and sort of slides off and sits there being Tempting. Very Tempting ... Dad says Mum should not eat the crispy skin of a chicken cos it is full of fat and Mum will get an Art Attack. But it is yummy. And Dads wot won't take off their headphones and come and help have got No. Right. to lecture Mums on wot they should eat. Hespecially when they can't see ... and luffly boykins who do Oppy-Mistic Hanging Around can find themselves on the fabumazing end of a-bit-for-me and a-bit-for-Worzel chicken skin ... Mum hasn't yet worked out how I do know about Oppy-Mistic Hanging Around when the chicken is getting carved. Or wot actual happens to the chicken skin when Dad does the carving. Or why Dad *usually* offers to carve the chicken. But she will ... heventually...

october 19

Fings have gotted much betterer on the sleeping and snoring and numb feet front. Mum's Feery and Hi-Pothy-Sneeze do seem to be actual right, and Dad is Not. Allowed. To. Get. Too. Tired. He's not sure how he does feel about this, to be quite actual honest: his snoring does not bother *him* a-very-tall, he reckons. Apart from Mum poking him and then yelling at him and then being really, wheely grumpy the next day. That bit bothers him in the hear-hole department a lot and a lot, so he has decided to go along with Mum's going-to-bed earlier hinstructions. And if he doesn't want to do that fing, then he does sleepering on the sofa where he can snore-to-his-art's-content. And not get poked and roared at by Mum.

All of this is exerlent noos for a luffly boykin. Without Dad waking up Mum every ten minutes, I has binned able to reclaim my spot on the bed. I was forced to sleep downstairs last year when Harry was staying with us, and sleeping on the bed isn't somefing I are willing to give up on a permy-nant basis. Sleeping on the big bed might not have binned part of my dopping contract, but it has becomed a nabbit. Dad reckons I are far betterer at serting my rights than he is.

But since Mum has worked out Dad's snoring solution, she isn't waking up all the time and realising her feets have gonned numb. And heven betterer, she has started to fink of the hole feet numbing stuff as A Good Fing: she has remembered that, for a long very time, I was not prepared to do touching anyboddedy or anyfing when I was sleeping, and would have hystericals and jump off the bed and run away big-time badly down the stairs if I did heven fink somboddedy had touched me when I was asleep. Now, though, I will do resting my head on Mum's feets or even her knees if I are quite actual sure they won't do moving around too actual much. Mum has decided that this is progress and has decided to do looking-on-the-bright-side. Dad reckons Mum can never find her car keys, but is the World's Hexpert at finding silver linings.

october 22

There's a fuge orange round fing on the kitchen table. And it isn't Frank ... Mum says it's a pumpkin and it's a kind of vegetababble wot she is going to carve up so that it looks like a scary monster. I don't know why she is bothering to be quite actual honest, as it does already seem to be quite scary a-very-nuff. Tonight, every single cat that lives here has come in through the cat flap, jumped onto the table, had a quite actual too close hencounter with the Orry-frying Fing, shot back out through the cat flap, bumped into another cat on the way in, had a squabble, and then disappeared off into the night. I aren't hexactly sure there is much more she can do to it to have a biggerer himpact.

Dad says he isn't scared of it. Yet. It really depends on how he is very hexpected to get hinvolved with it. He says carving a pumpkin is hard blinking work, and Mum is never going to get through the skin with the kitchen knives. He's wondering how much of his shed he can fit in the back of his car so that when he goes to work, Mum doesn't get frustrated and start trying to chop it up with his power tools. And then either chop off her hand or worse, splatter pumpkin hinsides all over the kitchen and his hexpensive tools. Dad says he's stucked between a rock and a Nard Place, and even though it's not very actual comfy, he's more than happy to stay right there for as long as he can get away with it. The rock, he reckons, is the pumpkin or hand-splattering in the kitchen. The Nard Place is Mum getting inside the pumpkin, carving it up and then deciding to make somefing that he will be actual hexpected to eat. That's when it will get scary ...

october 23

The fuge pumpkin is still on the kitchen table, and all of the cats have decided that as it hasn't made any sudden moves, it is probababbly not as actual scary as they did first fink. And also they is quite very hungry, which is more himportant than being terry-fried or hoffended.

Mum has asked Dad again to get-her-started with the pumpkin, and I

For THE QUITE ~~very~~ actual LOVE of Worzel Wooface

do really, wheely hope he does do this fing before Mum remembers about the chopper in the back of the drawer in the kitchen. The last time the fuge ginger boyman was home, he did use that chopper to attack a big bit of venison that Mum needed cut up into smallerer bits for my dinner. And he did flailing about and making all sorts of Hi-yaaaa! noises which were hessential for persuading the chopper not to be so blunt. But the chopper is even more actual blunt now, and the neighbours already fink Mum is bonkers-crazy. I fink her Hi-yaaa! noises will be even louderer than the fuge ginger boyman's were ...

october 25

Mostyn, Gandhi's litter bruvver, has gonned missing – again. His Mum says she finks she knows what the problem is, but she doesn't know how to solve it, and does Mum have any hideas? She has dunned some hinvestigating around where she do live and a noo cat has moved into her street, and the two cats have gotted into lots of harguments. It's hupsetting and frightening for Mostyn, which is why he keeps bogging off, his Mum finks.

Round where we live the cats do sometimes squabble. Mum counted up yesterday, and there are fourteen cats wot do live within yellering distance of my backdoor. Mostly they do keep out of each other's way, and the honly place there is ever troubles is down the little path that runs alongside the back gardens. Mostly, I do solving these harguments by doing exerlent hobserving, and then when I can't resist any actual more, I do woofing and commenting on the action. If it's a really, wheely hexciting squabble, Pip and Merlin do commenting from their end of the path as-very-well. Then both the cats wot are showing off their bestest puffing-up-like-a-balloon skills do have a shock and become more hinterested in working out where the woofing is coming from ... and do runnering away in different directions. Then they sit down and start to do washering and grooming themselves, and pretending nuffink happened, and that they were hinnocent bystanders.

Mum did have lots of not necessary hinstructions for Mostyn's Mum, about putting up posters and phoning up all the vets and asking the neighbours if they had seen him. And also checkering that none of them had haccidentally decided to do doppering him without realising he already had a home and a famberly who love him. But Mostyn's Mum had already dunned all that. And a lot of crying and not sleeping and worrying. So all Mum could offer was some kind don't-worry-too-much-yet words but they was probababbly as wasted on Mostyn's Mum as they woulda binned on my Mum.

Mostyn's Mum has heven been down to the lorry park to let them know he's missing again; she is just hoping that he is okay and hasn't been hitted by a car.

october 26

Mum and Dad have reached a compo-wise about the pumpkin. Hunfortunately, I do seem to be the compri-wise, and now I has got Complaints to the Management.

Dad says he will do getting hinside the pumpkin so long as Mum doesn't

130

try to put it hinside of him. Any of it. Mum says that's fine, she's going to Give. It. To. Worzel. She says I are not to worry; I'll honly be having the fresh bits that come out when she carves it, not the rest of it that has sat on the doorstep being very hadmired cos that will be old and mouldy and ick-yuck.

I are possibibbly, probababbly, definitely actual very sure I does not want to be used as a compo-wise: I are sure I does not like pumpkin. First of-very-all, it's a vegetababble, and second of-very-all, it isn't chicken wings. It does seem that my hopinion about pumpkin eating is being quite actual hignored ...

october 27

... and so is Mum and Dad's compo-wise. A compo-wise is *supposed* to be when noboddedy is actual happy with a hagreement but they will do livering with it. In our house, though, a compo-wise is when Mum says just a-very-bout anyfing to get Dad to do somefing, and then fibs. Or does somefing completely actual different and if-he-doesn't-know-it-can't-hurt-him.

Today, Mum has binned using a noisy machine that squashes and squishes vegetababbles really, wheely fast until they turn into a mushy, soupy mess, which Mum says is called a sauce. She's going to use it to make some dinners with pasta, then feed it to Dad and he'll-never-know. Mum says pumpkin is about to be added to the fuge list of vegetababbles that Dad doesn't eat that he's been eatering for years. And that's why he hasn't got ill with scurfy or fat or had a art attack.

I did also have a go at squashing and squishing pumpkin but I did not use the noisy machine. I did use the hard bony bit on the side of my face that I do find works just as-very-well. Honly I did it much actual quieter than Mum and her noisy squishing machine. And also in the sitting room, all over the sofa.

But I won't be eatering the pumpkin: I aren't stoopid a-very-nuff to compo-wise with Mum ... hunlike Dad, who has just actual hannounced that his pasta dinner was fabumazing, heven if the sitting room is a vision in Norange, and what-were-you-doing-when-Worzel-was-painting-with-pumpkin?

october 28

After all the pumpkin painting yesterday, last night I did have to have a bath. Bathing Worzel is Dad's job, because I do have to be lifted in and out of the bath and persuaded that staying there is a Good. Hidea. It isn't; it's a very rubbish hidea. The bestest hidea is to not get in it a-very-tall, but Dad and I do puttering-up-with-it. It's called being Stow-ickle and we is both hexperts at this fing.

Last night's bath did actual hinvolve more compo-wising: Dad said he was honly going to bath Worzel if Mum did somefing about my pumpkin-painting in the sitting room, so I was hopeful that I might actual havoid the bath. As it turns out, though, this bath was not a normal bath to get rid of the squished pumpkin that was welded to my face and all over my collar. Happarently, I would have had to have a bath whether or not I was norange because tomorrow somefing Himportant and Hexciting is happening! I are going to London to see Auntie Beverley because she needs to borrow me to be a prop. I aren't sure how I do feel about being borrowed or what being a prop is all about, but Mum says she will be actual with me the hole time, and she-wouldn't-miss-it-for-the-world. All will be revealed tomorrow, but after all this bathing, I are hexpecting Auntie Beverley to reveal some flipping cheese Big-Time-Badly.

October 30

I are home from my trip to London, and I are quite blinking actual knackered. My day started too very early, and did first of all hinvolve a train trip to London and a fuge train station called Liverpool Street, where a man kept yellering THE-NEXT-RAIN over and over again. I are quite actual sure he had to keep yellering this because nobodddedy was listening to him, and also because he kept shoutering his words up a helefant's bottom and actual hexpecting them to still make sense when they came out of his trunk. At least that's how it did sound to me. Heventually, we did make it out of the train station, and Mum did decide that as we had plenty of time, we would walk to Covent Garden. Mum asked a couple of peoples to help her get her phone the right way up so she could do walking in the right direction. None of the peoples were very keen to help with this, though, and all they wanted to say was it's-a-very-long-way and why-don't-you-get-the-tube?

Now, I do realise that not everybodddedy can do living in the countryside like my famberly does, but two miles is *not* a long way a-very-tall; that's a small-to-naverage walk for a luffly boykin, with henergy lefted over to do plenty of playing with Kite afterwards. London peoples honly seem to use their legs for making sure their bags don't fall over, I do fink. Perhaps if they did all do spreading out a bit and not trying to live on actual top of each other they might find that their legs worked for somefing else. Like walking, which was Mum's plan because I had binned a super boykin on the train, and going on a hescalator and then on a tube might be a bit too actual much for Worzel. And all the signs said Dogs Must Be Carried and you're-having-a-laugh, haccording to Mum.

Our walk through London was very actual hinteresting. There were lots and lots of lampposts to sniff, and I did leaving plenty of messages to let the London doggies know I had binned to visit. Nuffink very actual much had changed from my last trip to London, though, and everybodddedy did stare at Mum like she was walking a donkey through the middle of the City.

Once we did get to Covent Garden, there was a nice café with some Not-Grass-Really-Carpet for me to lie on whilst Mum and Auntie Beverley did their talking, when All. Was. Revealed. I did discover that it was my Himportant Work to do standing on a white circle and be booful and dorable whilst Auntie Beverley did talking. And then it would go on the telly. Not on the telly news but for a programme with a funny man.

Fortunately, I was considered a Very Himportant Dog, and I was allowed to do checkering out the room where the filming was happening, and have-a-go at standing on the white circle to make sure I would not have hystericals and either use my brakes or bog-off big-time-badly. As there was a lot and a lot of cheese on offer in return for standing on the big white circle, I did decide I would do this fing.

Later on, when it came to doing-the-telly-filming, I was actual shocked that instead of the room being hempty like it had binned when we did the practicing, it was stuffed full of peoples. Fundreds of them. But Auntie Beverley is Hexperienced With Dogs, and she did exerlent hencouraging me without being worried. All the fundreds of people had binned tolded that they must not do clapping or making any sudden noises, but a few of them did forgetting this fing and there were a lot and a lot of ahs and awws. Ahs and awws do not bother me a-very-tall, and soon I was quite actual relaxed and henjoying myself. I did perfickly! And after it was all finished, I got more bitsa cheese and luffly strokes from Mum who had watched the Hole Fing from behind the camera, where she did crying like she always does when one of our famberly does somefing fabumazing.

By the time we had dunned our telly-filming, it was quite actual dark, so I did get to hexperience a Noo Bitta Transport. I had a go at being a Londoner in a taxi, which was very actual hentertaining, and I did henjoy lookering out of the windows. I was not allowed to do sittering on the taxi seats so I did standing up the hole way, wot was quite very bouncy, just like being on a boat. Mum reckons the honly bits of transport I has not dunned now is a hairy-plane and a Kay-Ball Car, and she's not sure I would like neither of them fings.

Today, I has spended all day in bed. Mum reckons fame-mouse peoples honly get out of bed for fousands of pounds, so I has decided that now I are a fame-mouse telly dog, I reckon I can get away with that fing today. But I don't fink the fame-mouse peoples stay in bed because of the monies: being fame-mouse is quite very actual flipping hexhausting – and I honly had to do it for one day!

october 31

Mum has putted her carved pumpkin on the back step where it can be very actual hadmired by ... noboddedy! Cos our garden does have a fuge fence around it and noboddedy can see in. Mum says that is Not. The. Point. She is very actual pleased with it, and that's The. Point. Tonight, she will do puttering an ickle candle in it so that the face she has had a go at carving will glow and look heven scarier. It's all part of Halloween, what is trad-ishon.

This afternoon I did spot the pumpkin on the back step ... and I did peeing on it. And making it very actual soggy. Cos it didn't smell of me and it's noo, and it's what I do to anyfing noo in the garden.

It might not have a fancy name like Halloween but ... it's my trad-ishon and I'm stickering with it. After all, that's also The. Point ...

November

November 1

Mostyn has still not gone home and his Mum is worried sick. She's losted some weight and got some sore patches on her eyes from worrying. She's looked everywhere and is starting to fink that she will never see him again. Mum has reminded her about Mouse and her hincredible Bogging-Off-For-Five-Years, and says there is always lots of hope. By the end of the phone talking, I do fink Mostyn's Mum did feel a bit betterer, and she says she'll murder-him-when-he-gets-home.

November 2

Dad says that Mum was very good with Mostyn's Mum yesterday but she's-a-right-blinking-Hippy-Crit. A Hippy-Crit, I has discovered, is someone who is exerlent at saying one fing and then doing somefing the complete actual opposite. And Mum is being hippy-critty-call as whenever one of our cats goes missing she has complete hystericals and can't sleep. Then she makes sure that noboddedy else in our village can sleep neither by going out late at night and callering their names, and then getting up four hours later and doing some more name-callering. And also very hexpecting Dad to do joining in.

Each of our cats has a different length of time that they is allowed to be gone from the house. Frank is not allowed to go anywhere a-very-tall, wot is his own actual fault. He is always on the kitchen table so he is always actual hexpected to *be* on the kitchen table. He's a bit like the cooker: he lives in the kitchen, and if when you went downstairs first fing in the morning and there was a fuge, cooker-shaped gap, you would either fink you'd gone mad or that someboddedy had robber-dobbed you. Frank has to be in the kitchen or appear as soon as the larder door creaks. One day he won't do that fing and then everyboddedy will fink he is deaded. Which he had better actual blinking be, to be quite actual honest, as it wouldn't hoccur to anyboddedy that he'd got actual shutted in the garage or anyfing more sensibibble or hopeful.

Of all the cats that are not allowed to get lost, Mouse is the most likely to cause hystericals. Just finking about Mouse going missing again makes Mum hupset. Mouse has to be in the kitchen in the morning, ready to argue with Mum about whether she is actual allowed to switch on the kettle before she gets out the cats' breakfast. What she does during the rest of the day is anyboddedy's guess, and I would not advise noboddedy to spend too much time wondering what Mouse does during the day, or what is going on hinside her head: hinside Mouse's head is the sort of place you could visit and never find your way out of.

Mabel is allowed to miss either breakfast or dinner but no more. She has a nabbit of falling asleep in the shed and not hearing Mum calling. And actual sometimes, she will decide that there is Somefing Odd in the garden that she is quite sure has been putted there to hoffend her. Like the pumpkin wot I peed on. Or a milk bottle. Or heven a leaf that is moving in a different direction to the one Mabel wants it to. She isn't wot you'd call fussy about the fings that freak her out. Most cats, if they see somefing they aren't sure about, will make sure it isn't alive and can't hurt them, and then stalk past it, pretendering it doesn't hexist. But Mabel takes it to a hole nother level: she pretends *she* doesn't hexist, and hides until Mum comes to rescue her and remind her that she is still alive ... and probababbly quite hungry.

Gandhi is allowed to be missing for a hole day and a night without noboddedy worrying. He is probababbly the honly normal cat we do have, and he does an exerlent job of showing all the other cats what they could be doing with their actual lives if they wasn't so weird. Or fat. Or stoopid. Gandhi isn't any of these fings and, haccording to Dad, everyfing about him has get-up-and-go. He always does lookering smart and neat; his tail always sticks straight up in the air, and he does winding himself around your feet and sleeping neatly on hooman's laps, like all the cats on the telly do do. Mum reckons he's just like that really, wheely perfick kid at school who always has his pencils sharpened and never forgets his PE kit. He makes the rest of us look like very actual scruffy and lazy, but as he was borned and bred here, he also does gettering away with being perfick because we is all quite actual proud of him.

But none of the cats, even Gandhi, who is perfickly capababble of lookering after himself, would be actual allowed to be missing for as long as Mostyn has now binned gone, and we is all starting to fink that he has gotted himself into terribibble troubles, and perhaps heven been hit by a car. It's the-helefant-in-the-room, Dad says, that noboddedy's talking about, and fortunately, Mum did not mention it to Mostyn's Mum. There is No. Helefant. in any of the rooms of my house, I are pleased to say, and I hasn't had to do much checkering, either, cos helefants are fuge. But there is a tents feeling of worries and wobbles wot are bigger than any helefant.

November 3

I has got a noo toy. It's a Nedgehog and my friends, Sue and Nick, did bring it for me. It makes the strangest noise I has ever, hever heard, and it did make me want to put my head on one side to do hearing it betterer. Happarently, this is called Ned Cocking, and it is dorable, and now I has learned some very quite hinteresting fings about hoomans that I do fink might be Very Himportant for other doggies, hespecially hogpig Labradors.

Hoomans will do anyfing, and I do mean habsolutely anyfing, if you do cock your head sideways. It's like chicken wings for Mums and Dads. They will lie on the floor with you, they will squeak the toy hendlessly and in different and hinteresting ways, if there is heven a chance of you doing this Ned Cocking. Forget sit, or down. They is very nuffink compared to Ned Cocking.

Today, Mum needed lots of distractions from worrying about Mostyn, and so has spended all afternoon lying on the floor with her camera, squeaking the toy and very hencouraging me to do Ned Cocking. And each time I has dunned it, not honly have I been gived a bita cheese, she has dunned squeezing the toy again. And taking photos. I has had more bitsa cheese than I know wot to actual do with!

And then, when Dad came home, he was greeted by a very quite hexcited Mum, full of come-and-look-at-Worzel words, and we gotted to do it all over again. Cos heven rufty-tufty Dads are quite actual powerless when they see a vision of a luffly boykin Ned Cocking. They do turn into mushy, soppy hidiots and ... I fink I has just hinvented another way of clicker training hoomans.

November 5
****BESTEST NOOS!****

After nearly two hole weeks, Mostyn has binned found. He was taken to the same vet he got tooked to last time, where they did scanning his microchip and then phoning his Mum. Again. It was all very actual hembarrassing for Mostyn's Mum but she doesn't care: he is alive and he is back home.

She says she is going to shut him in and try to work out what to do actual next, but for tonight, she's just going to give him his favourite dinner and have cuddles.

November 7

It's going to be Dad's birfday at the weekend. Mum says she hasn't got a clue wot he would like for his birfday because she's got him just a-very-bout everyfing he does need over the years that she can hafford. Which honly leaves the fings that cost fousands of monies that we don't have. Like a brand noo boat. So she is going to buy him a big bar of white chocolate, which she finks is disgustering and wouldn't-eat-if-it-was-the-last-fing-in-the-house. That way it's a proper present for Dad, and not a present for Mum with Dad's name written on it.

Today I has dunned some fortful finking about my Dad and now that I has knowed him for over four actual years, I do fink I might nearly be a hexpert at hunderstanding him.

****FORTS ABOUT DAD****

- In the morning, the most himportant fing about Dad is to remember where he does keep his gentleman bits. This is much lower down than you might fink, and a luffly boykin like Worzel Wooface can ruin Dad's day before it has even-blinking-started-you-great-oaf, if he does forgetting that fing

- Dad is very quite good at being horganised and getting himself out of the house in about three seconds. Mum says this is cos she has to do everyfing in the mornings. Dad says it isn't: he doesn't leave his car keys in a different place every day and have to spend half-an-hour hunting for them. I shall leave you to do makering up your own minds ...

- Everyfing wot happens in my house is Dad's exerlent hidea. If it wasn't Dad's hidea, he did hagree to it when he was watching the football. Or playing a Confuser game with his headphones jammed over his Hearholes

- It is Dad's himportant work to turn off lights. Every day. Every. Blinking. Day. he says. I don't fink anyboddedy else here is qually-fried to do it

- Very actual hoccasionally, Dad will put-his-foot-down. Happarently, it's another way of saying no. It works halmost as well as when he uses the word 'no.' It just takes a bit longer to turn into yes

- Snoring is how Dads get all the noises out of their mouths that they can't say during the day, cos the Mums use all the air for their talking

- In the evening, when my Dad comes home, he does actual hexpect Worzel to get off the bed and come downstairs to say hello. If Dad has had a Bad Day, and he is knackered, it is hessential that I do ask for lots and lots of pats and strokes. Then he can sit on the stairs and pretend to be having a special time with me whilst he gets his breath back

- My Dad is very actual brilliant at fixing fings. There is nuffink he can't fix. Apart from singing Teddy Bears and World Peace. Not even to stop Mum being sad at the television news. And I are quite very actual sure he could fix the singing Teddy Bear; he just isn't that stoopid

- Dads don't have to be your actual bi-oh-lodge-ickle Father, I has discovered. My Dad did dopping me when I was an ickle Worzel Wooface, and as far as I are concerned, that is quite actual good enuff for me

- Some peoples don't have Dads. Mum doesn't have one, and she says she did just fine, but if you does have a Dad, I do hope he is kind and gentle and rubbish at saying no. Just like mine ...

November 8

Happarently, dogs can get named after food. Good food, not lettuce food, but sausages. There is such a fing as being a sausage dog, and I would like to actual be one. The previously ginger one says I can't be a sausage dog as my legs are far too very long. But she would like one a lot and a lot, and then she would be able to go on Sausage Dog Walks that are very quite pop-oo-la at the moment. Mum reckons the previously ginger one should have a go at doing some Lurcher walks with Worzel Wooface before she does commit to the hidea of having a dog of her actual own. Sausage dogs might be short but they is still proper dogs called Dachshunds, and they does still have to be trained and stooded outside with for their wees and poos in the pouring rain, none of which she has time to do if the previously ginger one gets unwell.

In the meantime, I would like to know how I can get myself renamed after cheese. Dad says I should stop finking these forts cos if I was named after wot I last eated I would be a cat-poo-dog, and could I please stop-diggering-it-up, as now Mum knows he did let me out in the front garden again last night for a wee and he is in the Dog House. Big-time-badly.

November 9

Mostyn's Mum did phone us last night and she has fort of a solution but she doesn't like it. In very fact, she hates it and she was crying, but she says she has to put Mostyn first and fink of his happiness.

She asked Mum if Mostyn can come back to live with our famberly. And Mum did say yes, himmediately. When Mostyn and his other bruvver and sisters were getting their famberlies choosed for them, one of the fings Mum did make everyboddedy promise was that if they could no longer do looking after their cat, then they had to contact Mum and we would have them back straightaway. I aren't sure Dad did know about these promise words because I fink he would probababbly put-his-foot-down at the fort of living with ten cats. How-very-ever, very

For THE QUITE ^very actual LOVE of worzel wooface

all the other kittens are very actual happy with their famberlies so I do fink that is hunlikely to happen. Having said that, Mostyn was actual happy until about three months ago, and you can never tell what will happen in the future. That's why peoples shouldn't do making too many puppies or kittens, and should be responsibibble and make sure they have always got space and room to look after them for the rest of their lives, Mum says.

And then she did practising these words in the mirror a couple of times before she phoned Dad and tolded him that, tomorrow, as well as gettering a big bar of white chocolate for his birfday, he'll be getting a cat ...

November 10

Mum is flipping between being habsolutely frilled to bits about Mostyn and feeling very actual bad and guilty because she does know that his poor Mum's heart has broked to actual pieces. Dad is being stow-ickle about the hole fing. He does hunderstand about the back-up-for-life promise that Mum did actual hagree with Mostyn's Mum, and anyway, Mostyn is booful and dorable. And Dad's rubbish at saying no.

So he has decided to remind Mum that he said she shouldn't go lookering for another cat and that one would find us. I has gotted habsolutely No. Memberry. of when Dad did say this fing, but Mum's not harguing. If this is what it takes for him to decide it was somehow his hidea, that's just fine with her: she has gotted hexactly wot she does want without any fuss or harguments.

November 11

Hugh-Stan we have a fuge problem. I does not know who Hugh-Stan is but that is wot everyboddedy does say when they does not know wot to actual do, so I is saying it right very now. Mostyn is hidentical to Gandhi. And I mean Hi-Dent-I-Cal. There is no way of telling them apart. Currently, all the hoomans have gived up calling either of the cats Mostyn or Gandhi and instead are calling them both Witch-Won-Are-Yoo?

At the moment, the hoomans working out Witch-Won-Are-Yoo? is not a Pri-Orry-Tree, and everyboddedy is trying to make sure that the cats wot already live here do behaving nicely. The cats are not struggling wth working out which one of the Witch-Won-Are-Yoos? is which because they can do smelling the difference, but they do need some time and space to do deciding Wots-Swat.

Wots-Swat is a very complercated cat-deciding fing where they do work out whether to hignore, hobject or have hystericals about the noo hinterloper cat, and it is somefing I are Not. Gettering. Hinvolved. With. Mum finks I are being very quite actual wise, and I has got to hagree with her. Fings are complercated a-very-nuff hinside the cats' heads at the moment without me hinterfering. I has got a orribble feeling that their complercated feelings might suddenly come out of their claws if I appear at the wrong moment, so I are refusering to come downstairs to eat my dinner.

But it is all actual worth it. Dad says I are a lily-livered chicken so I do seem to have got myself a foody name after-very-all! Happarently, it isn't a compliment but, as he did bring my dinner up to the landing and do sittering with me whilst I ate it, I are quite actual living with it.

November 12

The Witch-Won-Are-Yoos? have decided to become a Boy Band. There has binned a lot and a lot of singing the songs of their hancestors. I don't fink it's going to catch on or become pop-oo-la. At the moment the hole wailing in Arm-and-Knee is a work-in-progress as they don't seem to be singing the same song, even though they is singing at the same time. And there don't seem to be any words. All the hoomans do keep asking them to shut-up-and-get-over-it.

November 13

Hoomans do have to live by lots of rules, and luffly boykins have to remember quite a few as-very-well. But for cats, as far as I can hunderstand there is honly one rule, and it is that all wees and poos must be dunned outside. Today, Mouse has decided that seeing as Mostyn has to wee and poo in a litter tray as he isn't allowed outside for two weeks, that rule no longer applies to her. Mum says that's marvellous and now she'll have retrain Mouse again in a fortnight. Again again. She's lost count now of How. Many. Times. She's. Done. It. Now.

Tonight, Mostyn will be sleeping in a crate. That way, Mum says, she can put the litter tray inside it and away from lazy-cats-and-disgustering-dogs. I are currently in The Nile about the hole litter tray fing. I fort I might be able to actual get away with being a little secret house fairy that does tidying up when everyboddedy is asleep, but it does seem my hefforts are not happreciated. I fink I probababbly knew they wouldn't be, but it was worth a try ...

November 14

Dear Mum

Mouse says she couldn't find the litter tray last night. So she used your laundry basket hunder the kitchen table.

From your luffly boykin

Worzel Wooface

Pee-Ess: Dad says there was No. Poo. in the laundry basket when he went to work this morning.

Pee-Pee-Ess: That's halmost, possibibbly, probababbly not true ...

November 15

Gandhi is Not. A. Happy. Cat. In very fact, he is flipping furious about Mostyn, and being a blinking-bully-leave-him-alone. Any hopes we did have about him gettering-over-it are fading very actual fast. Usually, Gandhi is very quite keen to be out and about, doing his exerlent showing off at being a perfick, normal cat, and just appearing in the morning and at night to have his dinner. Hoccasionally, Gandhi brings in his own dinner wot is usually actual quite hinteresting and get-away-Worzel and please-come-and-help-me-get-rid-of-this-now and flipping-heck-it's-still-alive.

But now Gandhi has decided he needs to Sert His Rights about this being His. House. and he isn't going anywhere. He is trying to be very In.

Charge. of everywhere, which is himpossibibble, Mum says. She's tried. And everywhere Mostyn wants to be, Gandhi finks is his himportant special place. Dad finks he's got a blinking cheek, and he's been treating-the-house-like-a-No-Tell for the past two years.

November 16

It would make more actual sense if Frank or Mouse were doing Complaints to the Management about Mostyn. They do spend most of their time indoors but haven't made any hobjections a-very-tall, and Mabel is being her usual the-world-is-going-to-end self as always, so there's no real change there. I are starting to fink that there are some benefits to being a less-than-perfick-cat. It's quieter, for a very start. Or it could be that they are too stoopid to realise, too lazy to care, or too hysterical to hang around long a-very-nuff to cause troubles.

Wotever it is, Gandhi is not being his usual perfick self. Or maybe he is. Maybe being furious about a hinterloper is normal cat behaviour? And I has got to say, Mostyn isn't helping himself a-very-tall. He is not fighting back but he isn't hignoring Gandhi, neither. Dad reckons that when he got terry-fried at his old house and did runnering away, it did haffect him Big. Time. Badly. and so he is feeling and acting like a victim which is rewarding all of Gandhi's being-a-bully hideas and forts.

Mum has decided that she needs some advice from someboddedy wot does know more about wot goes on hinside cats' heads than she does, so tomorrow she is going to phone Uncle Boris at the vets to see if he has any cunning plans, because she would really, wheely like Mostyn and Gandhi to be happy together, and this isn't how she did imagine it would work out with them. Ever since Gipsy did do dying, she has binned wittering on and on about being a Man Down. Or a Cat down. Dad reckons he isn't anyfing Down. Apart from down the boat. Hiding. Huntil the Witch-Won-Are-Yoos? decide to stop howling at each other.

November 17

Uncle Boris has tolded Mum to try somefing called scent-swapping with Mostyn and Gandhi. It hinvolves taking some of Mostyn's bedding out of his crate and putting it where Gandhi chooses to sleep most of the time. And vicey-versy. And in the meantime, we should do keeping them apart for a few days until they has tried just living with each other's smell for a bit.

He has also dunned suggesting that we do get somefing called a De-Phew-Sir wot zaps happy smells that cats like around the house. Dad says it'll make all the cats become stones. Or somefing like that. I has got No. Hidea. wot he is talking about, but as far as I are concerned, so long as Mostyn and Gandhi stop squabbling they can act like blinking boulders for all I do care.

November 19

Gandhi and Mostyn have stopped singing with each other! Well, currently they has had no hopportunities to do practicing together because they is not actual allowed to meet. All the other cats have decided that Mostyn is okay. And now fings have calmed down a very actual bit, we has been able to spot some differences between Mostyn and Gandhi. Gandhi is fatter than Mostyn, and his eyes are closer together. That is not hexactly plite about Gandhi, but then he is not wot you'd call pop-oo-la at the moment, so noboddedy is finking a lot and a lot about his forts and feelings. I do fink this is Most Hunfair to my Onorary Lurcher Gandhi: usually, he is cheerful and positive about everyfing, and all this being-a-bully to Mostyn is probababbly because he is quite very actual hupset.

Now that fings are a bit calmer, I has binned able to meet Mostyn properly rather than do exerlent pretending he doesn't hexist on the way to the garden. And I are pleased to say that Mostyn does fink I are very nuffink to worry about, and has gived me a small nose bump and even a face rub. I did remember my manners and managed not to get very too hexcited about this. Mum says she gives up trying to hunderstand animals any blinking more: not honly are Mostyn and Gandhi both cats, they is brothers, and litter brothers at that. So how is it that they can't bear to be in the same room as each other, but both of them can tolly-rate a dirty great Lurcher? I has decided to hignore the dirty comment and put it down to Mum being in hurgent need of wine, she says.

November 20

Fings have tooked a different cultural turn in our house. Now that boy band singing has binned abandoned, Mouse has decided that she would like to have a go at some artistic doings. And I are pleased to say that it does not actual hinvolve litter trays. Mouse has taken to having a go at recreating fame-mouse works of art in real life. Happarently, it is called Performance Art, and she has started doing an exerlent himitiation of the Silent Scream. She does open her mouth and make all the right faces for a fuge great meee-owww ... but nuffink comes out.

I aren't very actual sure if Mouse knows that no noise is coming out. She doesn't look surprised at the Silent Scream or have another go with the sound turned up. Perhaps she can hear it in her head, or she is so used to the sound that comes out that she's kinda tuned it out. Mum is trying to remember the last time she actual heard Mouse make a noise out of her mouth, and keeps rattling a few bits of kibble in a cup to see if it works at all. So far, Mouse has managed a few squeaks but mainly it's all Silent Scream. As far as I are concerned, Mouse can carry on Silent Screaming for as long as she wants. After the racket we've had for the past week it's a fuge very actual relief to have some Not Noise to worry about.

November 21

Today, Mostyn and Gandhi were actual allowed to meet hunder supervision. It did go very quite well, and there was no howling or being-a-bully or terry-fried victim doing. It was all quite actual calm. And progress.

Tonight, though, Mostyn decided that he did not want to go into his crate to sleep and did act like a toddler that didn't want to go back in the pushchair, clinging onto the sides of the crate and trying to spread himself out in as many different ways as he could so he wouldn't do fitting through the door. Mum reckons this could also be called progress because he isn't wanting to hide away all the time, and hopefully, in a couple of days, he will be able to go outside now he knows who puts food in his bowl. It's going to be a bit of a risk letting Mostyn out of the house because there is always a chance that he will Bog-Off-Big-Time-Badly like he used to do at his old house, or decide that Gandhi and our house is just as terry-frying and run away from here. But, on the other hand, there is a fuge lot more space outside so Mostyn and Gandhi will have a lot and a lot more chances to havoid each other.

Mum says she isn't pushing her luck, though, and tonight, whether Mostyn likes it or actual not, he Has. To. Go. Back. In. His. Crate.

November 22 (2am)

Dear Mum

I aren't sure how to tell you this, seeing as it is the middle of the night, and you is fast asleep, but I has just been into the sitting room and now I do know why you did have such a fuge problem persuading Mostyn to go into the crate.

You will be pleased to know he has gived up with his Complaints to the Management and has gonned to sleep. Well, Gandhi has. Cos it's Gandhi in the crate. And Not. Mostyn. Mostyn has worked out how to do the cat flap and is outside in the garden having a good sniff about.

From your luffly boykin

Worzel Wooface

November 22 (3am)

Dear Mum

Mostyn has dunned jumping over the fence and I can't see where he has gonned. I did trying to give you a biff to wake you up but you did tell me to lie-down-and-stop-being-hidiotic. This is now Your. Problem.

From your luffly boykin

Worzel Wooface

November 22 (7am)

I has got No. Hidea. how this hasn't ended in actual very disaster but holy-hell-you're-not-Gandhi has just strolled in through the back door when Mum did callering the cats for their breakfast. The real Gandhi is Not. Amused. and has showed his fury at being hunlawfully detained overnight by scattering the hentire contents of the litter tray through the bars of the crate, and hurling it as far as he could all over the carpet. Gandhi and Mostyn's scent-swapping is now swapped all over the sitting room floor ...

November 23

It's binned a Nellavaweek, Mum says. Nellavaweeks are when everyfing does seem like Too. Much. to actual cope with. So now Mum is drinking Sherry with Kite's Mum, which is like drinking wine. Honly faster, she says.

Then toys gotted thrown outta prams and Dad had to go and scrape Mum off the concrete. Neither of those fings actual happened – it's all somefing to do with mettyfours. And Dad was quite glad Mum didn't have a pram or toys to throw out of it, and he didn't need a shovel cos that woulda binned painful. As it was it was all very actual loud.

And then Mum tried to help cook dinner which, with all the Fast Wine she'd drunked, wasn't wot you'd call a success so she cut her fumb really, wheely badly, and then did a lot of crying, wot wasn't about her fumb a-very-tall, happarently, but it was a good blinking hexcuse.

I fink it's all quite very actual serious and worrying cos Dad is being, well ... helpful. There's binned halmost no tutting or running away to the harbour a-very-tall. Fortunately, it's Sunday tomorrow and hopefully the Nellavaweek will be over. And me and Mum HAND Dad are all going for a fuge long walk down at the beach. And we will halmost definitely, probababbly, possibibbly, maybe come back.

November 24

The previously ginger one wants to leave home which is wot is causing all the Nellavaweek and hupset. And the trouble is there is nowhere for her to actual go. There aren't any small flats or safe places, but she wants to be hindependent so she is going to go, anyway. Mum is frantic and worried and doesn't know if the previously ginger one will actual be able to stay alive. It's all very quite worrying and frightening, and no-one can persuade her to slow down or make some plans.

The Men-Tall Elf people are not helping a-very-tall. They are completely useless, Mum says, and she cannot be actual trusted to be in the same room as any of them at the moment. Fortunately, I has not binned put in the same room as them neither cos if Mum was wobbly and furious, I are not sure how I would actual behave. Dad says seeing as I has got that hoption, staying out of it is very actual wise. He does wish he had that choice ...

November 25

There was a feasant squawkering in the field on the other side of the Hedge. And I Did Recall. I does deserve a Meddle. Made of Cheese. About the same size as the actual Moon! Mum says I are a luffly boykin and she's glad she's got me around at the moment or she might go actual bonkers mad with worry and hupset.

November 27

The previously ginger one has lefted home. I do not know where she has gonned, and Mum says the place is Orrendous and she-wouldn't-let-a-dog-stay-there. I are quite actual glad to hear that she would not put a dog in an Orrendous place but it does not sound like it is very much fun for the previously ginger one neither. But happarently, she is Adam Ant that she is Not. Coming. Home.

November 28

Mostyn, I do fink, is brighter-than-the-naverage cat. Or he is brighter-than-the-naverage cat wot lives here, at least. Mostyn and Gandhi have dunned reaching some sort of hagreement. So long as Mostyn doesn't use the cat flap or the back door to do coming in and out, Gandhi will tolly-rate Mostyn's hexistence. Just. So. Long. as he doesn't breathe too loudly. It's that kind of truce.

None of the other cats is very pleased with Gandhi about this; they has all decided that Mostyn is quite actual bearababble and heventually they will decide to be friends with him. I fink it would happen a lot and a lot faster if Gandhi would stop blinking remindering them all that Mostyn is a Noo Cat. But Gandhi, Dad finks, has got a Nin-feery-Ority Complex cos Mostyn does look like him, honly slimmerer and a lot and a lot less cross. Today, Gandhi did do his hole blow-up-like-a-puffer-fish and dance around on his tippy-toes and wail the Songs of his Hancestors. So Mostyn ranned away from the back door, over the fence, and out of the garden.

Gandhi was very actual pleased with his hachievement. And Mum did a lot and a lot of sighing and worrying that fings might be betterer but they isn't perfick yet. But then, there was a tiny mee-ow at the Front Door. Our other door. The door that none of the other cats seem to know actual hexists. So whilst Gandhi was guarding the cat flap and the other cats watched him and wondered what dance he was planning to do next, and also did hoping that he would stay down at ground-level this time, and mainly not throw the sugar bowl all over the floor, and most himportantly Not Drag Them Into It ... Mostyn got letted in the front door by Mum and has strolled hupstairs. To his favourite warm cosy place in the bathroom on top of the towels. And Gandhi is STILL guarding the cat flap waiting to pounce ...

November 30

I don't fink there has binned any fuge harguments wot have made the previously ginger one want to leave home. There hasn't been any shoutering or banging doors. There hasn't heven binned any real hatmosfears for me to worry about. Not anyfing sudden, anyway, and it doesn't feel like the hatmosfear has gonned along with the previously ginger one. In very fact, I fink now it has gotted worse. Mum is tents and stressed and Dad is saying halmost nuffink. It should feel quieter now there is one less hooman here but it actual isn't; there is a fuge buzzing Not. Noise. that is halmost louderer than the normal chaos of my famberly being busy. It's like the hole house is hinfected with Mouse's Silent Scream.

But no matter what Mum says to her, the previously ginger one is still Not. Coming. Home.

December

December 1

According to the poxy guv'ment I aren't Senty-Ant. I are! I flipping-well-blinking-are Senty-Ant. Here is a list of some of the Senty-Anting I has dunned in the last 24 hours –

- Joyful when Dad camed home
- Hinterested in sniffing out the noo smells in the garden lefted by a Nedgehog
- Hungry when I was waiting for my dinner
- A ickle bit cross when Gandhi tried to pinch my chicken wing
- Scared when there was a loud bang on the telly
- Shocked when Dad flinged the bed covers on top of me when he got too hot in the middle of the night
- Tired cos Mum couldn't sleep and kepted waking me up with her uppy-downy tossing and turnering
- Cold when Mum made me go outside this morning for a wee
- Hexcited when the postman banged on the door
- Somefing-along-the-lines of 'oops' when I was caughted trying to eat some cat poo I founded in the garden

Now, I are quite very actual prepared to be the first to hadmit, that I don't feel or fink about all the fings wot hoomans do do. To be quite actual honest, I are very actual glad about that fing ... all the finking and feeling hoomans do is all very hexhausting, and if I did all that much finking and feeling there would not be a-very-nuff time to do playing or snoozing ... BUT – quite very actual BUT – I does have feelings, and so does every-other-blinking hanimal in the hole wide world. I are Senty-Ant and so is they.

December 2

It's all about Being A Burden, and that is why the previously ginger one has gonned to live somewhere else. And also about monies and no-one having a-very-nuff of it to go round, plus a dollop of not-living-in-the-middle-of-nowhere, plus a fuge lot of Mum panicking and worrying about her All. The. Time.

Mum says the previously ginger one might not want to be a burden but moving out hasn't actual helped; it's made fings worse. Trying to keep someboddedy safe who has a Men-Tall Elf causing chaos hinside their head is hexhausting and frustrating, and mostly a fankless task that can drive almost anyboddedy into gettering their own Men-Tall Elf. And now Mum has got to try and keep the previously ginger one safe from 15 miles away. Which might be possibibble if the Men-Tall Elf services in this area were any blinking good. But despite all the telly yellering and new peoples being very In. Charge. of it all now, it's going to be actual flipping ages until the changes make any very actual difference to the peoples who need help. And now all the talk is about somefing called Early Hintervention. Dad wants to know how they are going to help the hole generation of peoples wot should have gotted early hintervention but didn't. Who are now much iller than they needed to be. It feels like they have all been abandoned as a lost cause.

OVER-CATTED

Dad says he's been over-catted
By Gandhi and Frank tonight
One cat and one lap is wot it should be
And Frank isn't wot you'd call light

Frank's a fuge and great lump of a pussycat
And he will not sit himself down
He keeps shoving himself up Dad's nose
 And hendlessly wriggling around

Gandhi's much better at lap-doing
He's the stillest cat you could meet
Or it could be he can't blinking move
Cos Frank is squashing his feet

Them cats they has finally settled
All is peaceful and calm, as you see
But Dad's not feeling comfy ...
... he can't breathe and he needs a wee

December 3
All of a actual sudden, Mostyn has decided that Gandhi is his worstest henemy again. This morning, Mostyn has binned very quite grumpy, and there has been a lot and a lot of growling and hissing. Hespecially at Gandhi. Just recently, fings have binned much calmer, and there has been hardly any singing songs of Hancestors wot was getting a bit boring and like-a-broken-record and could-they-please-give-it-a-rest?. Fortunately, Gandhi has decided not to do joining in with Mostyn's growling and hissing. He has more himportant fings to do finking about: a fuge bowl of yummy wet cat food that Mum has gotted out to distract him and the other cats whilst she does working out wot the Heckington Stanley is wrong with Mostyn.

December 4
There is quite very actual definitely Somefing. Wrong. With. Mostyn. Yesterday, he did decide that lying in my bed in the Hoffice was wot he needed to do, and I are pleased to say that I was an exerlent stow-ickle boykin and letted him do gettering away with this fing. To be quite actual honest, the look on his face and the way he kept diggering his claws into the cushions was a-very-nuff to convince me.

I aren't sure if Mostyn choosed my bed because it looked comfy or because it is right next to Mum's chair in the Hoffice, and he wanted her to hear the actual very strange sounds he was making. Mum has knowed Mostyn since he was actual borned. And it was she who did turning him hupside down and peering at what was going on hunder his tail when he was honly a few days old, so she does know he is a boy cat. Which is a good job because if she hadn't habsolutely Knowed. That. Fing. she woulda swored he was trying to have a baby. He did sound like a car trying to start on a frosty morning, and it was Orrendous, even from the top of the stairs.

Cats that do live in our house are not allowed to sound like cars trying to start, so Mum decided that he needed to go to see Boris the Vet hurgently. She fort that perhaps he had dunned twisting himself or gotted a habcess from a previous squabble. Boris fort that, too, huntil he did give Mostyn a bit of a squeeze. And then Mostyn had a go at rearranging Boris the Vet's face. Which would have binned a bit of a shame; Boris is quite very And Some. And dishies. Mum says.

It does turn out that Mostyn has got a completely Blocked-Up-Bladder. Currently, we does not know why. We just know that he really, wheely needs to get it hunblocked before he does hexploding. And stop growling and trying to remove people's noses. To do that he is going to have to stay with Boris at Wangford Vets.

Once he has gotted hunblocked we will know wot the problem is. And Dad will try to work out how much overtime he needs to do ...

December 5

The previously ginger one might have lefted home but halmost none of her stuff has. Her bedroom looks like the Hapocolips with added mould. At first, Mum fort the previously ginger one would come runnering back after a couple of days, saying sorry-I've-been-a-plonker-what's-for-tea and can-I-borrow-some-monies? Dad did a lot and a lot of hinsistering Mum did do nuffink about the mess and the yuck, and the what-on-very-earth-did-that-used-to-be? Otherwise, the previously ginger one might actual hannounce that she was leaving home once every ten days just to havoid having to Voover. Mum fort about that and decided that it was possibibbly actual true, and also muttered a lot of words about her being more-like-Dad-than-he-realises.

But the previously ginger one has binned gone nearly a week now, and today Mum decided that she couldn't very actual havoid going in there any more. And she could really, wheely do with somefing to distract her from worrying about Mostyn, on top of wondering whether the previously ginger one was eating anyfing. It seemed like a perfick time, and since there was a danger that the mould and the yuck might start coming out and hinvading the rest of the house, like the smell was starting to, probababbly overdue.

At first, all of the clearing up and folding hendless clothes and trying to match up earrings did make Mum hupset. And then cross about the fings that had binned trodded into the ground and ruined. And then very actual hexhausted after having a fuge fight with the bed to move it out of the way to get to the stuff that was jammed hunderneath it, and Stopping the somefing-

147

beginning-with-B Drawers shutting properly. With added kicks and shoves. But when she had finished, she came and sat on the top of the stairs and I did get some fortful finking strokes. Those strokes are the ones that hoomans don't actual realise they are doing: like a meddy-tation with hands hinstead of saying Omm over and over again. Mum doesn't do meddy-tation: she doesn't have time, she says, and muttering be-the-change-you-want-to-be hendlessly is a great way of not stacking the dishwasher and doing-the-jobs-you-need-to-do-before-Dad-comes-home. But quite very sometimes, she does do fortful finking strokes that are halmost the same, I do fink.

Heventually, after a a hole hour of fortful finking strokes, Mum stood up and said 'Right.' Then she marched down the stairs, out of the house ... and left.

Left left. Not left as in Right. Mum might be a bit bonkers-crazy and distractered by all the worry about the previously ginger one and Mostyn, but she didn't go marching right-left-right-left all the way to the harbour. She went in the car. But she said her be-back-soon words so I are halmost, possibibbly, probababbly sure that there is nuffink a-very-tall for me to worry about ... I may have binned left but everyfing is going to be alright.

December 6

It was quite very actual late when Mum and Dad gotted home last night. They had been to have their dinner in a restaurant. Some famberlies do go out for special meals when they has got somefing to celly-brate. In very general, mine do not do that fing. They go out for fancy food when they do need to concentrate and do talking, as far away from their phones and Uff the Confuser and Hemails as they can get.

Now that they have gotted home, Mum keeps saying stuff about madness and doing-the-same-stuff-again-and-again. And then hoping fings will be different. Which it actual won't. It doesn't matter How. Many. Times. I try to fit into the ickle bed Pip lefted here, I still cannot get all of me in it. Even though I really, wheely want to and try halmost every day. And Dad has binned trying to persuade Mum to put her keys on the shelf on the front of the larder door ever since he builded it so she doesn't lose them all the time. But she always forgets and he's forever moaning about it. You'd fink he'd have gived up by now ...

But last night's doing-the-same-stuff talk at the restaurant wasn't about Pip's bed, or keys, or larder shelves: it was all about the previously ginger one. And it turns out that although she has dunned it all the completely actual wrong way round and caused quite a fuge lot of hupset and worry, some fings have gotted better for everyboddedy since the previously ginger one lefted home. There is more cups in the kitchen, for a very start, Dad started to say, but Mum gave him A. Look. and tolded him to stop being a fortless hidiot and concentrate on the himportant fings. The previously ginger one has told Mum and Dad that she doesn't feel like Mum is panicking around her the hole time now, and that when Mum panics it makes her very actual anxious. And then Mum gets worried and the hole worry-and-hawful-feelings hescalate, just like when Mouse and Mabel have a squabble, and then Frank haccidentally comes in through the cat flap and lands on top of them. Honly worserer.

And although the previously ginger one is living in a Orrendous place and probababbly isn't eatering a-very-nuff, or remembering to do any washing, somefing positive has binned discovered.

So now, Mum and Dad do very hagree with her. And wotever she says now, she's Not. Coming. Home. In very fact, coming home would be a Bad Fing, and it's himportant that everyboddedy works together to make sure she doesn't have to come back here to live. It was honly when Mum started looking at the mess and seeing the broken fings and realising how much time the previously ginger one had spended lying in bed being poorly-sick, or gettering really, wheely hunwell and needing to be tooked to Opital by the green men with flashing lights, that Mum did do realising that if she came home, the same fings would keep on happening. Because Mum and Dad, and most himportantly, the previously ginger one, have tried and tried to make fings different and better but it hasn't worked.

It's going to be scary and a fuge risk in lots of ways, but if fings stay as they always have, then the same fing is going to keep on happening. And that would be Hutter Madness.

December 7

Mostyn has dunned a wee. Well, that isn't quite actual haccurate ... Boris did sticking a tube up his gentleman bits wot did a wee for him. So now he is not going to hexplode.

I are mainly finking about the Mostyn not hexploding good noos bit, and not about the Sticking A Tube up hinside his gentleman bits part. Dad says he is also finking hard about not finking about that too very much. Mum reckons it's a boy fing, and tried to do tellering Dad all about cat hanatomy, but he just winced a lot and we both does actual hagree that sometimes Mum is Not. Sensitive. to our forts and feelings a-very-nuff. And also that cats aren't built anyfing like wot we did fink, and why are all their gentleman bits in the wrong blinking place and pointing the wrong way? And surely that can't be comfy. A-very-tall.

December 8

Mostyn has a big hinfection and he could get all blocked up again. He has got to carry on staying with Boris-the-Vet and with his not-comfy-not-finking-about-it tube for a couple more days until he can do a wee without the tube ... and without getting blocked up again.

In the meantime, Dad and me has actual decided that Uff the Confuser does have Too Much Hinformation. And so does Mum, but mainly could she stop showing us pictures of cat willies ...

December 10

I aren't sure if Mum is fabumazing or a pain-in-the-wotsit. Dad says he knows that feeling Honly. Too. Well. Either way, Mostyn is HOME! Mum has binned trusted a-very-nuff to take care of Mostyn at our house. Probababbly so she doesn't keep phoning up Wangford Vets all the time to find out how he is feeling. He camed home last night With. His. Tube. that was actual quite hinteresting and a bit actual scary for Mum, but she did manage to carry him carefully to his crate without catching it on anyfing or hurting him. The crate is now up in the previously ginger one's old room where it is all very clean and quiet, and where none of the other cats will disturb him, so he can do recovering without gettering stressed and tents.

Everyboddedy here is feeling quite actual sorry and worried about Mostyn. He's had a lot and a lot of terribibble luck, and he does not deserve to be so hunwell, just as fings were starting to look brighter for him.

December 11

Today, Mostyn has had his tube tooked out, and I are pleased to say that fings seem to be working very quite okay in the wee department. Hunfortunately, he has got to wear a hateful buster collar for another couple of days so that he does not do too much hinterfering with his gentleman bits. Mostyn is Not. Pleased. He tried to rub up against Mum today for some haffection, and got quite actual frustrated that all he could rub against was a lump of hard plastic.

December 12

The previously ginger one is coming home! But honly to visit and get horganised and honly for a week. Just a-very-nuff time to sort out all of her fings, and then she is going to move to a place that isn't Orrendous, near where we do have some more famberly but in a different part of the country. It's not too very far from where the fuge ginger boyman lives, and most himportantly, the Men-Tall Elf services are about the very best that there is in the Hole of Ingerland.

Mum and Dad do have very mixed forts about all of this. On the one hand, the previously ginger one is 20 and does need to start to do lookering after herself. And Mum really, wheely does need to let her do this fing without stepping in and rescuing her every time somefing goes a bit wrong. Otherwise, Dad reckons, the previously ginger one will never learn that clean clothes don't magically appear in wardrobes by themselves. Mum was very surprised when Dad said these words; she was quite very actual sure that Dad fort this fing, too.

But it will still be a fuge worry, and it will all take a lot and a lot of gettering used to and everyboddedy will miss her so actual much, heven me. I has spended most of my life since I came to live here following Mum up the stairs to make sure the previously ginger one is still actual breathing, and not doing that any more is going to take a lot of gettering used to.

December 13

I are quite actual not pleased to say that this week I has not been a Pri-Orry-Tree. Also, there is not a-very-nuff doors in my house, and the ones that we do have are getting shutted with me on the Wrong. Side. Of. Them. Hand I has binned hexpected to be patient and tolly-rant of this fing. Wot I does fink is most hunfair. And not to my liking a-very-tall.

It's all to do with Mostyn and his poorly bladder, and making sure he does not get at all actual stressed and tents.

Stress, Boris-the-Vet says, is one of the reasons why cats do get bladder hinfections that then make them block up. And if it has happened once, it can happen again, which is somefing none of us want. Hespecially, Dad's bank account. So it is very, very actual hessential that Mostyn doesn't get stressed. Not honly will it hurt him a quite fuge lot, but there won't be any monies left for cheese.

December 14

Kite's Mum has dragged my Mum off to have a go at making a Reef for Crispmas. It's a kind of flower harrangement wot hangs on the outside of the front door to do scaring away the postman. In case he brings any letters asking for money, I fink, like the pumpkin at Halloween that was supposed to do scaring away ghosts. Mum's reef, Dad reckons, is a reflection-of-its-maker-and-the-place-it-was-made. Mum was nearly very pleased with Dad's forts about her reef ... huntil she did realise that he meant it looked like it had binned made by a mad woman who'd drunked too actual much mulled wine in the pub where she made it.

December 15

Mum and Dad have talked and fort a-very-lot, but they has comed to realise that there is no way it would be safe or kind for Mostyn to stay being our cat. He really, wheely needs to be a honly cat. He got frightened into runnering away at his last home, and he has hobviously carried all those being scared feelings to our house. Gandhi and Mostyn have just a-very-bout gotted used to tolly-rating each other, but none of us are that very conferdent that fings will stay that way. Mostyn being ill and poorly seems to have brought back Gandhi's feelings of being a big-bully again, probababbly because Mostyn is being a bit actual cross and sulkering. A couple of times Mum or Dad have caughted Gandhi trying to sneak into the previously ginger one's room, heven though there is nuffink in there apart from Mostyn. And none of us did fink that he was going in there to be comforting or to make plite sickbed conversation like hoomans do when they visit their famberly in Opital.

The trouble is, noboddedy is actual hinterested in making Mostyn their cat. He is having to eat hexpensive special food to protect his bladder, and Mum can't make any promises that it won't get blocked up again. Mum doesn't know any squillionaires, and noboddedy can very afford to take the risk. A couple of peoples have said that Mum should do rehoming Gandhi. But Gandhi has lived here for four years, so that just seems very actual wrong. Hand totally Hawful. And by the somefing-beginning-with-B words that came out of Mum's mouth, it isn't going to happen ...

December 16

It is a quite very actual true fact that Mum Talks. Too. Much. She'll talk to anyboddedy about anyfing. But very sometimes, those anyboddedys do turn out to be someboddedys, and today, Mum has remembered that she knows a Very Special Person who might be able to help with Mostyn. His name is Niall and, according to Mum, he is the best Animal Warden in the Hole Wide World.

If anyboddedy can help find a solution for Mostyn it will be Niall, she reckons.

December 18

You-flippin-Reeky! Mum has spoken to Niall, and he does know a fabumazing cat rescue that has just had a fuge load of monies lefted to it by someboddedy kind and not alive any more. It's called a Leg-and-See and it's somefing everyboddedy should fink about doing. Not necessarily the fuge bit if they has famberly, but an ickle bit. Anyway, Niall has had a chat with the rescue and because he is the best Animal Warden in the hole wide world who has helped out the rescue in the past, Mostyn is going to be found a home through them. And they will use some of the Leg-and-See to make sure he gets his special hexpensive food when he goes to his forever home, Hand, if his bladder does go wrong again, they will make sure he gets the right treatment. But most himportantly, they will find him a home as a honly cat where he can be the gentle and cuddly boy he does want to be. And not have to worry about other cats frightening him ever, hever again. And Mum won't have to go and live in a cardboard box or do selling one of her kid's knees, which is wot she was muttering about yesterday.

December 19

Mostyn has gonned on his hadventure to his noo life. He left today with all his fings from this house that do smell of him. Mum says it feels wrong and odd. But the right fing at the same time. Usually, she is the one doing rescuing other peoples' pets when fings go wrong, and lookering after foster doggies. She did never, hever fink she would have to ask a rescue to help her with a pet of ours.

None of the haminals in this house is perfick. We're all a bit odd and have our 'issues.' Mum couldn't have tried harderer to try and make it work, Dad says. But what me and the cats can't do is disappear to make the pet-free home that Mostyn so very actual needs. So Mum's dunned the honly fair fing she can fink of. Dad keeps reminding Mum that Niall is the best Animal Warden in the hole wide world, and he has promised to keep us up to date with Mostyn so we know he is safe and happy. Now, we just have to do actual living with it.

December 20

It's honly five days until Crispmas, but apart from the scary reef on the front door, there is no other signs of Crispmas anywhere in our house. Mum says that she doesn't feel like puttering up a tree this year. I fink this is probababbly wise and not just because of all the dramaticals with Mostyn and the previously ginger one. Mouse is having to start learning how to be house trained All. Over. Again. and the last fing she needs is an outdoor tree coming hindoors and confuddling her heven more.

Dad says we're going to go down to Corny Wall to see Granddad for a few days over Crispmas and keep him comp-knee. It will be his first one without Granny Wendy and he's going to have a proper Crispmas Dinner and not somefing that goes ding.

December 21

Just recently, when we has binned on a walk and metted some nice and hinteresting peoples, Mum has binned asked Harold Izzy? A lot. And then when Mum does tell them they does all say the same fing: Izzy Reely? I does find this all very quite confuddling. I aren't Harold. Or Reely. And I are definitely not Izzy. I are Worzel and I are four years old. It isn't old but it isn't a puppy, which is the Hole Point, happarently. Cos peoples do seem to fink that my hello-luffly-boykin greetings are quite, well, puppy-ish. I do fink that is quite a very actual hachievement when you is the same size as a table. I did fink I might have to be Hoffended and do Complaints to the Management about people finking I was a baby-puppykins, but Dad says it is all very okay. It's all about Hattitude to Life and you're-as-old-as-you-fink-you-are.

Mum finks she feels about 83. Dad says he's 27 in his head, which is why he did spend an hour playing with me on the marshes, and being funny and hentertaining and making Mum smile and giggle more than she has dunned for the past two months. But now Dad's 27-year-old-head has caughted up with his 54-year-old-knees. And he's quite very glad Mum finks it's so blinking funny, but can she give-him-a-hand or he's never going to get back to the car.

December 22

The silent screaming wot has hinvaded our house over the past few weeks has quite very actual gonned. Well, Mouse is still doing her best at it but it's starting to be sometimes silent and sometimes its normal wailing self. Now that Mostyn has gonned, we don't need the De-phew-sirs plugged in any more, and yesterday, Dad began to wonder whether there was somefing in them that made Mouse's mee-ow disappear.

I aren't actual sure but tonight, I are really, wishing they was still plugged in. An Ignormous Queen Wasp did decide to fly into the house, lookering for somewhere warm to sleep for the winter. Dad, in his somefing-beginning-with-B-Cider-fuelled-wisdom, fort he would try and swat it with a noospaper. Right above Mum's head. I don't fink I has ever, hever hearded anyfing with an art beat scream so loudly as Mum did do. And then she snatched the noospaper off Dad. I fink she mighta swatted Dad with it if she hadn't seen me having hystericals at the screaming noises. So she did exerlent pretending to be calm so as not to actual frighten me any actual more. It's not the same as real being calm (I are exerlent at spotting the difference between the two), but I did happreciate the heffort. In the meantime, me and Mum are hiding in the bedroom until Dad has actual hencouraged the Queen Wasp to find somewhere else to sleep. And Mum has stoled Dad's cider: she needs it far more than he does at the moment.

December 23

I are off to visit Granddad in Corny Wall today. And his hills. Mainly his hills, I do actual hexpect. And his stream. And Auntie Sue's cats, and then eat her Crispmas dinner! I are really, wheely lookering forward to being in Corny Wall, it was fabuamazing last time and there were lots and lots of hinteresting smells. And fings to roll in.

For THE QUITE very actual LOVE of Worzel Wooface

On the way to Corny Wall, we will be going to visit the previously ginger one where she is now living, and do giving her some presents and also going out for a meal. Mum says we will be seeing the fuge ginger boyman on the way back. He has gonned to have Crispmas with the Kat Lady and her famberly.

December 25

Sometimes, you can make all the bestest plans and have the best hideas. And fings still do go actual wrong. Yesterday evening, Auntie Sue, who was In. Charge. of cookering Crispmas Dinner, began to feel hunwell. And by ten in the evening, so did Uncle Dave. This morning, neither of them can actual do moving without feeling like they is going to fall over and not get up again. Both of them has tooked to their beds and say Crispmas is actual very cancelled. Mum is now very In. Charge. of finding somefing for us to eat at Granddad's house because all of our Crispmas dinner is actual stucked with Auntie Sue and Uncle Dave. No-one is feeling brave a-very-nuff to risk going to rescue it from their house. Heven Dad, who hates food wot goes ding.

Mum says it's quite very actual okay, though. She's found some food that doesn't have to go ding in the bottom of Granddad's freezer, and we will be having sausage and chips for tea. Dad finks this is fabumazing: for the first time ever, hever, noboddedy is going to try to hinsist he eats Brussels sprouts on Crispmas Day.

December 26

Auntie Sue and Uncle Dave are still really, wheely poorly-sick. They have got all the medicines they do need, and the Most Himportant Fing is that everyboddedy stays away so that *they* don't get sick, too. It's too late for that, though. Dad wented to visit them on Crispmas Eve and he says he is starting to feel a bit hot and shivery at the same time. Like someboddedy has chucked a bucket of freezing water all over him whilst he is standing in a hoven. He says it is our himportant work to run away from Corny Wall as quickerly as possibibble so he can do sick and dying in his own bed. But most himportantly so that he doesn't give it to Granddad who is ninety years old and doesn't need to catch flu. Granddad says he has had his flu jab so he should be okay. But then so did Auntie Sue, and she can't even get down her stairs at the moment ...

December 28

Dad has got The Leery-us. As well as The Lurgy. It's all part of being poorly-sick, Mum says. He keeps muttering about having to wait too actual long for his cheeseburger to cook, and wanting to complain to the manager of the takeaway shop. Honly we're not in a takeaway shop, he's lying in bed. And he doesn't like cheese ...

Mum says she Can't. Wait. For. This. Year. To. Be. Over. And she's not cookering Dad a cheeseburger no matter how much he yells. He can have a cuppatea if he can stay awake for long a-very-nuff to drink it, but mainly could he stop with his shoutering: it's scaring Worzel.

In betterer noos, Granddad is habsolutely fine. He has not caughted The

Lurgy or got The Leery-us. He says that, at 90, there isn't a-very-lot that hasn't tried to kill him. And this bug isn't going to get him neither.

December 29

Good noos! Dad has remembered that he doesn't like cheeseburgers and has stopped muttering about them in his sleep. And he has stayed awake for long a-very-nuff to drink a cuppatea. He wants to know how Mum is feeling, and does she fink she is going to get ill with The Lurgy?

Mum says she isn't going to be that lucky. There is No. Way. she'll ever get to lie down in a dark room for five days having someone take care of her every need. But it's all very okay, I do fink. If Mum does get The Leery-us, she does love cheese halmost as much as me, so there is nuffink to actual worry about.

December 31

Mum and Dad gotted to open their Crispmas presents today. Mum fort her favourite present was going to be the very quite hexpensive handbag Dad boughted for her. But then she opened a present from Kite's Mum: a quite very fabumazing picture that Kite's Mum did make her out of felt and bits of fluff. It's a picture of me, Worzel Wooface, looking dorable. And it made Mum cry. I was very actual halarmed by her wet and wobbly face: we've had far, far too actual many of those in our house this year.

But it turns out it was neither of them fings in the end. After all the present-opening and the wobbly and wet face-doing, Mum turned on Uff the Confuser to check her hemails and found the bestest present. Mostyn has found a new home. He's living as a honly cat with a very nice lady who does fink he is perfick. He hasn't had any problems with his bladder a-very-tall and he is quite very actual settled, safe and happy ...

THE END
(for now)

"Beautifully written, extremely witty & heart-warming – literally couldn't put it down – loved it!"
Dog Training Weekly

"I've actually met Worzel Wooface. I asked him what he thought of this book. From the look on his face, I guessed that he was quietly pleased, and so he should be"
Paul Heiney

"A real feel-good, laugh-out-loud book ... a must for all Lurcher owners"
Daily Express

***** AMAZON REVIEWS
• "A very actual fabumazing book!"
• "Bestest dog author ever is back!"
• "Worzel back, funnier and wiser"
• "Fabumazingly actual awesome!"
• "Am already looking forward to the next one!"

"Woo 2: required reading for anyone who has ever given their heart to a dog to tear"
Geelong Obedience Dog Club

"... also an exploration of human frailty, depression, achievement, courage strength and joy ... about unconditional love and an intimate acceptance that dogs think differently.
The World's Ace Woofer Does It Again! The third volume of the internationally acclaimed Lurcher Chronicles surpasses its predecessors. Live and laugh with Worzel! Learn to speak Lurcher, too!" – Geelong Obedience Dog Club

"Very very funny throughout, but also totally real and true to life ... beautiful pictures enhance the stories. Yet again an excellent paperback from Hubble & Hattie, and an absolute bargain at £9.99"
– *Dog Training Weekly*

***** AMAZON REVIEWS
"Worzel Wooface just gets better and better. I really cannot wait to read the next chapter in this truly wonderful dog's life. You feel part of his loving family. Sometimes funny; sometimes sad, but like all families full of ups, downs, love and laughter"

"... part of the [royalty] money is helping the shelter that Worzel actually came from. Recommend to all animal lovers"

"Cleverly observed; written with sensitivity and humour"

"Once started, it was so difficult to put down. Thank you, Worzel (and Catherine) for letting us share your ups and downs, and making us laugh"

For more info on Hubble and Hattie books please visit www.hubbleandhattie.com; email info@hubbleandhattie.com; tel 44 (0) 1305 260068 *prices subject to change

For children of all ages!

Children and dogs can be the best of friends, and good for each other in so many ways.

Sadly, though, dogs are often misunderstood by both adults and children, and their behaviour misread.

Worzel Says Hello! is the first in a new series of delightful educational and fabulously illustrated guides to understanding how dogs think, feel and behave, so that all children can have a wonderful relationship with the dogs in their lives, and all dogs can feel happy, safe, and loved.

Written by
Catherine Pickles

Hubble & Hattie

Illustrations by
Chantal Bourgonje

"The paw prints on the cover invite small hands to pick up this fabumazing jewel ... an accurate, aesthetic artistic, literary delight! Hubble & Hattie has set a formidable benchmark for canine educators opting for the popular picture storybook format" – Geelong Obedience Dog Club

***** AMAZON REVIEWS

"This book is an absolute delight. and the message it delivers in such a charming, gentle, down to earth and very sensible way. As for the illustrations ... suffice to say they did bring a lump to my throat"

"They will love it ... the 7-year-old will be reading it to the 2-year-old I am sure!"

"This lovely book turned up today, and me and my 6-year-old read it instantly, and fell in love with it. Simple wording for children and perfect illustrations"

"This book is excellent for teaching children and adults alike how to behave around dogs. The illustrations are exquisite"

"This book is brilliant. Easy to read for the small ones just learning, or for someone to read it to them"

"An excellent book. Gently shows children how to approach dogs, but also creates a lovely short story with a great rhythm to the words"

HH5160 • Hardback • 20.5x20.5cm • £6.99* • 32 pages • 49 colour pictures, inc 45 original illustrations
• ISBN: 978-1-787111-60-8 • UPC: 6-36847-01160-4

The second book in the Worzel Kids! series

Beautifully illustrated by Chantal Bourgonje, this is the story of a child and a dog going for a walk, told from the dog's point of view in an easy-reading rhythmical style.

As explained by Worzel, an enormous Lurcher with 'issues,' new experiences can be very scary for him, although he really, really does want to go on exciting walks. The original and engaging illustrations will delight both children and adults, as well as educate, showing them the secret language of dogs, and will help children learn how to walk dogs in a safe and responsible way so that it is a positive experience for everyone.

The information presented in this book is endorsed by dog trainers, parents and teachers alike, and is an essential addition to every classroom, library and child's bookshelf.

September 2018 publication • HH5292• Hardback • 20.5x20.5cm • £6.99 • 32 pages • 33 original colour illustrations • ISBN: 978-1-787112-92-6 • UPC: 6-36847-01292-2